human**instinct**

Also by Robert Winston

INFERTILITY
GETTING PREGNANT
MAKING BABIES
GENETIC MANIPULATION
THE IVF REVOLUTION
SUPERHUMAN (with Lori Oliwenstein)

human**instinct**

ROBERT WINSTON

BANTAM PRESS

LONDON · NEW YORK · TORONTO · SYDNEY · AUCKLAND

TRANSWORLD PUBLISHERS
61–63 Uxbridge Road, London W5 5SA
a division of The Random House Group Ltd

RANDOM HOUSE AUSTRALIA (PTY) LTD
20 Alfred Street, Milsons Point, Sydney,
New South Wales 2061, Australia

RANDOM HOUSE NEW ZEALAND LTD
18 Poland Road, Glenfield, Auckland 10, New Zealand

RANDOM HOUSE SOUTH AFRICA (PTY) LTD
Endulini, 5a Jubilee Road, Parktown 2193, South Africa

Published 2002 by Bantam Press
a division of Transworld Publishers

A catalogue record for this book is available
from the British Library.

ISBN 0593 05024X

Typeset by Phoenix Typesetting, Burley-in-Wharfedale, West Yorkshire.

Printed in Great Britain
by Clays Ltd, St Ives plc

3 5 7 9 10 8 6 4 2

To Tanya, Joel and Ben,
who seem to me to have
precisely the right human instincts

Contents

Acknowledgements

When I first started on the BBC 1 project *Human Instinct* I had only the most superficial interest in the subject, and matching expertise. Over subsequent months I have become so gripped by the notion that we are all trapped inside what is essentially a Stone Age mind that my interest at least has improved enormously. Fascination aside, this book has been written within the extreme pressures of a full diary and a varied employment as academic researcher, clinician, part-time politician and science film contributor – all, sadly, done less well than I would like. Whilst the deficiencies in this book are wholly mine, I have leant heavily on a number of people without whom this manuscript could not have been completed.

Foremost, I owe a great debt to Leo Singer. We first met and worked together when I was writing the book *Superhuman*. The remarkable intelligence, insight and lively interest he showed then have been surpassed by his input into this manuscript. I could not have completed it without his skills in researching and his excellent ability at distilling and editing; it has been the greatest pleasure to work with him.

I first met my friend Alison Dillon when working on the BBC series *The Human Body*. Her intelligence and scientific understanding were incredibly valuable then. Her ability to express complex issues in plain English, together with her broad science background, were extraordinarily helpful with both the style and content of this book.

The later chapters, and particularly the last one, concerning altruism and the nature of morality, cover a difficult and daunting area. I am indebted to Michael Pollak, whose philosophical input was invaluable and who suggested a number of rational drafting changes, and to discussions with Raphael Zarum, Robert Rabinowitz, Joanne Winston and Clive Freedman, who pointed me in useful directions. Both Lira and Joel Winston read the whole manuscript and made many valuable suggestions and improvements.

My own scientist colleagues have, as usual, been a tower of strength, warmth and support and have tolerated my distraction while working on this project. Firstly, I am deeply grateful to David Edwards, Chairman of my division at Imperial College, who is so supportive of the notion of science communicaton and who allowed me a mini-sabbatical. Particular thanks go to Kate Hardy, Debbie Taylor, Carol Redhead and Delyth Morgan, who have been the most wonderful and encouraging scientific collaborators, and for the friendship of Charlie Waters, who helped me with the drafting of the genome programme 'Threads of Life' which provided some background to this book, and for shared insights into the nature of human instinct. I thank Ray Owen for drawing my attention to material on instinct and race.

One of the greatest privileges in my hobby as a television presenter is to work in close collaboration with some of the most talented and creative production teams. My special gratitude goes to Phil Dolling and Jessica Cecil, who helped with much of the background material for this book and who made many useful suggestions about the manuscript. I must record my warmest fond admiration to three talented producers: Nastasha Bondy, who produced the programmes on survival and sex; Nigel Patterson, the programme on competition; Sam Starbuck, the last programme on altruism and morality. They have been an extraordinarily stimulating team with whom to work and I hope we meet again for other projects in the future. The research for the TV series, which provided much material for this book, was so competently done by Rebecca Chicot, helped by Dan Bays, Stefanie Kern, Laura Riley and our assistant producer, Nicola Cook. Thanks too to Mark Hedgecoe, producer on *Walking with Cavemen*, and my good friend Richard Dale (as usual a tower of strength); the many discussions we shared in the area of hominid development were engaging and valuable.

Maggie Pearlstine, my agent, and John Oates in her office have proved to be the usual enthusiastic and encouraging team. Without them this book would not have been started, let alone finished, and they have always been available for advice and support. My publishers, Bantam Press, have been wonderful – flexible with tight deadlines and highly encouraging. Particular thanks to Sally Gaminara for her amazing backing and unstinting commitment to the project, and to Daniel Balado for his copy-editing.

Lastly, as always, my thanks to Lira, for her tolerance, support and wisdom – and love.

The photograph of the male Argentine Lake Duck is reproduced with the kind permission of Kevin G. McCracken and first appeared in *Nature*, vol 413, 13 September 2001.

human**instinct**

Introduction

Man, the Most Inquisitive Animal

The reflector of the Arecibo Radio Telescope in Puerto Rico covers an area of eighteen acres. Its antennae, suspended 450 feet above the reflector, are so sensitive that the telescope could pick up the signal emitted from a mobile phone half a billion miles away. This instrument is perfectly suited to listening for signs of alien life.

Arecibo is the home of SETI, the search for extra-terrestrial intelligence. Every night this enormous dish probes likely star systems for signals that may indicate the existence of intelligent life. For each patch of sky the system crawls slowly along the radio dial, searching as many as two billion frequencies and spewing out vast amounts of data for analysis. Within this morass of electromagnetic noise the SETI team hope to find a pattern they cannot explain – signals that are not caused by interference from a telecommunications satellite, or the sweeping beam of radiation from a pulsar. SETI astronomers sit dreaming at screens through the Puerto Rican night, hoping for the magic moment when the system flashes up the message: 'Signal found – origin unknown'.

Is their mission futile? There are five million patches of sky to scrutinize, and the search is bedevilled by interference from an increasing number of man-made satellites. Even if there is intelligent life out there, the sceptics say, the chances of picking up an alien news bulletin amid so much random noise are pretty much zero. But even when all five million star systems within Arecibo's sights have been searched, analysed and cast aside as electromagnetic junk, the critics will not be able to claim victory. There will always be higher-definition, more sensitive telescopes able to focus on more distant star systems.

Whether or not SETI eventually meets with success, we cannot fail to be awed by the extraordinary possibility of intelligent life outside our own solar system. It has value not necessarily because intelligence is a gift from God, nor because we humans, as intelligent beings, have any inbuilt moral worth. Intelligence is precious in the same way as a flawless, deep blue diamond is precious: it is rare, beautiful, and appears to be not of this world.

Human consciousness is often portrayed as pure and all-knowing reason that can escape the confines of our biological existence. At the end of Stanley Kubrick's and Arthur C. Clarke's film *2001: A Space Odyssey*, the future of mankind is depicted by some kind of space-child in a ghostly floating bubble, watching over the earth and its people. But we are not free-floating souls, set free from bodily needs and experience. Our minds are rooted firmly in our evolutionary past. Our reason is bound up in layers of instinct, of prejudices, desires full of both selfishness and kindness, underpinned by a will to survive and a will to reproduce.

One interpretation of Kubrick's epic is that the black monoliths were left by alien visitors as a means to hasten the arrival of intelligent life. They appear to act as way-points in the evolution of our species: the first, which appears one day implanted in the soil of the African plains, encourages ape-men to pick up rocks and bones and use them as tools; when the second monolith is uncovered on the moon, we are impelled to explore the rest of the solar system, an event that marks the beginnings of our detachment from earth, like fledglings leaving a nest; the third gives birth to some kind of disembodied spirit, the 'space-child', an image of our future that is both absurdly fanciful and wholly terrifying.

The true story of human evolution is hardly less fantastic. In this book, I want to examine, from its origins in the east African savannah, the legacy left by those aliens who really were responsible for the birth of modern man: *Australopithecus*, *Homo habilis* and *Homo erectus*.

Most thinking people now fully accept the basic theory of evolution. Few people, except perhaps the creationists,* have much of a problem with the idea that humans descended from the apes and that apes themselves evolved from more primitive mammals. But while people have no problem with the idea that our general shape and structure are derived from other creatures, fewer consider, let alone accept, the psychological implications. *Homo sapiens* not only looks, moves and breathes like an ape, he also thinks like one. Not only do we have a Stone Age body, with many vestiges of our past, we also have a Stone Age mind. The pressures to which we have been exposed over millennia have left a mental and emotional legacy. Some of these emotions and reactions, derived from the species who were our ancestors, are unnecessary in a modern age, but these vestiges of a former existence are indelibly printed in our make-up.

So this book is essentially about the instincts that intrigue all of us. What, for example, drives a man like the mountaineer Joe Simpson, abandoned and dehydrated in dire weather with a broken leg and stuck deep in a mountain crevasse at over twenty thousand feet, to ignore his pain, above all refuse the comfort of sleep and spend three days crawling back to base-camp in a semi-conscious state? How is it that so many apparently happily married men fantasize about pretty, slim young women seen on a tube train, or risk the happiness of their partners, their children and their own peace of mind on a transient sexual experience? Why is it that so many thousands of people – mostly, but not exclusively, male – will spend their whole week entirely focused on whether Arsenal will win their next crucial football match against Manchester United, or whether the Arizona Diamondbacks can beat the New York Yankees in the baseball World Series, when they are overweight, smoke too much and have no sporting ability themselves? What stimulates that urge

* And sometimes, it seems to me from the huge correspondence I get from them, the creationists seem to be protesting so vigorously that they appear as if they are trying to convince themselves.

to press the pedal as hard as possible at traffic lights to make the fastest getaway? Why be surprised when fascists such as Monsieur Le Pen or Herr Haider are able to appeal successfully to the deeply instinctual racist views of so many people? How is it that humans show altruistic and empathetic behaviour when there seems no possible benefit in return? And how is it that so many people still hold religious views and profess belief in God when the notion of an all-powerful being is irrational?

This book examines these human instincts: survival, sexual drive, competition, aggression, altruism, our search for knowledge and our need for something more, perhaps the divine. Though we have travelled far in time, and moved away from the savannah which first nurtured us, it is there we must first look to find some of the answers.

The savannah

Five million years ago, our hominid ancestors climbed down from the trees in the thinning forests to try their luck on the savannah. They were forced by an encroaching Ice Age to adapt to a new environment, a place with fewer natural resources than the vegetation-rich forests and little physical protection from predators. Here, a slow and vicious drama of natural selection would be played out over two hundred thousand generations as the ape-men struggled to compete with animals that were faster, stronger, hardier, more poisonous and fundamentally more suited to the violence, mayhem and weather of savannah life.

We began this life on the savannah as *Australopithecus* with a brain the size of a chimpanzee's; over the next three million years it tripled in size. Our brain, and the mind it housed, appears to have been our secret weapon and the solution to the problem of survival. An increasingly complex mental architecture began to develop. Alongside an unprecedented expansion of the sheer *number* of brain cells (up to our present level of around one hundred billion nerve cells) came an increasingly sophisticated mind. We continued to evolve an array of instincts that went hand in hand with an extraordinary jump in learning, emotion and rationality.

We learned how to make and use tools. We discovered fire and the uses to which it could be put. We became curious about the wider world, and started to explore. We began to talk to one another, which allowed communal living to become more complex and more successful. Small groups of hungry hunter-gatherers could pool their resources, exchange crucial information about themselves, their environment and the availability of water, food or fuel. Larger groups became possible, held together by complex and emotional threads of co-operation and kinship. Increasing complexity and division of labour allowed us to put down roots, build civilizations and invent a rich cultural life.

As a species, we are not physically designed for large and anonymous cities, low-level stress, fast food, addictive drugs and the fracturing of communal life. Whoever invented nuclear weapons was not thinking about the ease with which we form alliances and turn to violence against our enemies. The pursuit of material wealth and status often involves the splintering of family units, and we have emotional needs and desires that are not always fulfilled. We were used to the gossip and intrigue that grew from a close-knit and interdependent group; now we must be content with *EastEnders*.

So there is tension between our Stone Age instincts and the stresses and strains imposed by post-industrial civilization. We are forced, as a species, to walk through life laden down with the genetic baggage of five million years of savannah psychology and the inherited traits that preceded the hominids.

Instinct and the genes

This genetic baggage is the subject of this book. But first, let me try to define what I mean by the word 'instinct'. What does Charles Darwin have to say on the matter? 'I will not attempt any definition of instinct,' he writes in *On the Origin of Species*. 'It would be easy to show that several distinct mental actions are commonly embraced by the term; but everyone understands what is meant when it is said that instinct impels the cuckoo to migrate and to lay her eggs in other birds' nests.'

So much for the father of evolutionary theory, you may say. But Darwin was right to point out that none of the characters one associates with the term 'instinct' will turn out to be cast-iron universals; there will always be exceptions. Of course, we should have a working definition, and it must turn on the distinction between the mind we are born with and the mind that is 'made', via learning, culture and socialization. Instinct, then, is essentially that part of our behaviour which is not learned. Nevertheless, our environment (and hence our learning) may have a powerful effect on the way our instincts are expressed. Instinct is those elements of human action, desire, reason and behaviour that are inherited, and those instincts which are specifically human are those that were honed during our time on the savannah. Nowadays we know a great deal more about inherited qualities than Darwin ever did – we know that they are transmitted through genes.

The completion of the sequencing of the human genome is a landmark in the history of the human sciences. One key figure in this international research programme was John Sulston, head of the UK Sanger Institute. There is an oblique and evocative portrait of Sulston, by British artist Marc Quinn, housed in the National Portrait Gallery in London. It is the gallery's first 'conceptual portrait'. It does not exhibit the face of its subject; instead, within a thick stainless-steel frame, rows of translucent beads sit suspended in agar jelly. Each bead contains millions of tangled strands of DNA, chemically amplified from a sample of Sulston's own genetic material, derived, incidentally, from a sample of his sperm.

It may be difficult to grasp the scale of the achievement of Sulston and his fellow researchers, such as the Americans Francis Collins and Eric Lander. In 1985, the various luminaries in the field first got together in Santa Cruz to discuss the idea and concluded that it was basically unfeasible. But in 1988 the project was given its official seal of approval by the American government. Some people likened the task to the moonshot, but for those involved it felt as though they had promised to put a man on Mars. That feeling quickly changed with rapid advances in DNA-sequencing technology and increasingly heavyweight computer firepower, and the Human Genome Project was completed in 2001. The publication is a list of three billion letters that represent the chemical rungs of the DNA double helix, denoted by the letters A, T, G and C. It runs to over six hundred thousand pages of A4, and in book form would take up 270 feet of shelving space.

Within this expanse of code lies the recipe for the development of the human body. While all of us (unless we have an identical twin) have minuscule variations, the vast majority of the chemical code is identical from person to person. That is why it makes sense to call it 'the' human genome. Roughly, one out of every thousand letters will be different from individual to individual. Within these differences lie the variations in human physiology, our hormonal balance, our tendencies to develop cancer, or to have blue eyes. But six hundred thousand pages of code is a big haystack in which to find these specific genes and link them to their 'phenotypic' effect – the manifestation of a gene's action in the organism.

The development of the human brain is determined largely by this genetic code. But the details of the cognitive workings of the brain are still largely a mystery. Investigation of its physical structure does not get us far, and much of our medical treatment is crude; it's as though I were given a screwdriver and asked to discover the details of a computer operating system by opening up the plastic case and examining the contents. The analogy is perhaps apt, for the human brain is not just like a computer – it *is* a computer, of sorts. Within its fast and flexible neural networks are data processors, memory caches, behavioural algorithms, logic-solving programmes, fast-response mechanisms, and inputs and outputs to the world outside. The best way to investigate this organ is not to open up the case, it is to boot up the software and watch it in action.

The vagaries of behaviour

The twentieth century claims impressive scientific achievements, beginning with Einstein's theories of special and general relativity, which audaciously challenged the orthodoxy of Newtonian physics. Relativity soon proved itself by accounting for observed peculiarities in Mercury's orbit and accurately predicting how starlight would appear to bend around the sun during a solar eclipse. The world was granted an extremely visible and violent confirmation of the accuracy of Einstein's theory when the USA dropped atom bombs on Nagasaki and Hiroshima.

But what about the human sciences, and attempts to cast light on 'human nature'? Anthropologists have returned from far-flung exotic islands, as well as inner-city ghettos, with wonderful stories of cultural oddities, bizarre and extraordinary rituals and beliefs. Some have attempted to construct grand and wide-ranging theories on how human nature and cultures across the world and across time are similar in some ways and different in others. But after a time their theories, such as functionalism or structuralism, fade away or are discredited. Professor Clifford Geertz, the highly respected anthropologist from Princeton University, is right to point out that anthropologists are not in the business of conducting laboratory experiments. The studies, he says, that attempt to show that the Oedipus complex appears to be 'backwards' among the Trobriand Islanders, or the theory that the Pueblo Indians are entirely non-aggressive, are interpretations rather than 'scientifically tested and approved' hypotheses.

Human behaviour is fickle and unpredictable. Just because Mr Pooter, the unassuming protagonist of *The Diary of a Nobody*, has caught the quarter-to-nine bus to the City every morning for twenty years does not mean that we can rely on him to catch it tomorrow; indeed, one does not have to wait long to find out that Mr Pooter *does* miss the bus, on account of an argument with a delivery boy. Mercury's orbit, on the other hand, is highly predictable, now that relativity has refined our Newtonian model of the movement of heavenly bodies, and unless it gets hit by a wayward asteroid or comet (an event that we could in any case predict) it will remain in this pattern until the sun burns itself out.

Day-to-day behavioural possibilities are endless. Imagine you're in a restaurant, eagerly awaiting your supper. You've missed lunch and you are like one of Pavlov's dogs, drooling at the thought of the grilled piece of sirloin you have ordered, medium rare. The waiter puts down a plate in front of you, but the steak is not just well done, it's charred black and tough as old boots. Your heart sinks. What do you do? Send it back? Make a scene? Point out to the waiter that it's overdone, and then eat it anyway? Wait what will seem like an *eternity* for a new steak? Or simply walk out of the restaurant and go to Burger King instead?

Now, imagine a particle of dust is floating through a room, illuminated by a shaft of sunlight streaming in through one window. The speck of dust will follow a particular track in three-dimensional space, its movement affected by a number of variables: draughts blowing through an open window, sunlight heating the air and creating thermals that lift the particle skywards, unpredictable eddies and vortices that spin off from these currents of air, and perhaps the extremely rare occurrence of the mite colliding with another airborne particle. Finally, there are the actions of gravity and air resistance, forces that act on the dust particle just like they act on any other object falling through the atmosphere. The track is more likely to be full of twists and turns, full of shifts one way and then the other. At any one point in time two or more of these forces may be acting simultaneously. Some of them, like the sunlight thermals and gravity, may act in opposite directions, and cancel each other out. Others will act in concert, speeding up the progress of the dust speck in one direction.

Just as the speck of dust is at the mercy of many forces, so is human behaviour. We are pushed and pulled in all directions by many different biological, cognitive and cultural forces. Some of these may oppose one another, and some may pull in the same direction. It is entirely possible that two instinctual tendencies may act at odds to each other. But that does not mean these forces cannot co-exist; it just means that the track through space is more difficult to understand. Just because we have one adaptive mechanism that promotes violence and another that promotes co-operation does not mean that our explanations of these forces are confused. We are pushed one way and pulled another, and our challenge is to try to disentangle the forces and explain their origins.

Chaos theory tells us that the track of the dust particle will be extremely difficult, if not impossible, to predict in advance. The reason for this is related to the butterfly effect, which theorizes that the flutter of the wing of a single butterfly in China can ultimately affect the course of a tropical storm in the Caribbean. Tiny shifts in initial conditions have a critical effect on the final outcome of a chaotic system, and this applies as much to human behaviour as it does to the physical world. It is impossible to model our behaviour because there are too many factors involved, each with the potential to wreak the same havoc as the flutter of the butterfly's wing. And there is an added complication: humans apparently have free will.

Explanation of a great deal of human behaviour is an extraordinarily complex process. It is the product of many different factors – instinctive, physiological, rational and emotional – and prediction becomes impossible. Laboratory experiments have shown that the time it takes for a person to react to a flash of light varies, and it varies according to no discernible pattern. Random output of this kind may be useful if one is being chased by a lion; we may change direction, jumping from the left to the right, in an unpredictable way, and we may be a little bit more likely to escape intact. Randomness, then, is an intrinsic part of our neural make-up. One evening you may eat the steak, and the next you may assault the waiter.

But we can make valid assumptions about how life must have been on the savannah and how early man might have reacted to many experiences. We know that the basic physical principles on this planet held fast. There was night and day, sunlight and rain, and variation in temperature. There were hills to climb, rivers to cross, and occasional droughts. We know that there was an array of major predators, mostly big cats, because we have found their bones. Hominids were in danger of being eaten. We know that there were herds of antelope, deer and other herbivorous mammals that were prey for the cats, and possibly for hominids too. Plant life would have offered nourishment, but it would also have harboured venomous insects as well as producing poisonous berries.

We can be confident that the ground rules of the mammalian lifestyle remained intact. Hominids on the savannah needed to eat, drink, keep warm at night and sleep. We aged. We went through puberty, and we had sex. Women fell pregnant and nursed their babies. They were physically slighter and weaker than the males. They were only able to have a relatively small number of children during their fertile period. Men were stronger, and there was no real limit to the number of women they could impregnate. We shall examine the implications of sex differences in later chapters, for they are extremely important to the story we are going to tell.

All of this we know with certainty. We know that hominids were familiar with death, of siblings, parents and children. There was disease and injury. The frail or injured were dependent on others for food and protection, otherwise they died. Infant mortality would have been high, life expectancy low.

We can also assume that the savannah grasslands were not a bottomless well of material resources. This Garden of Eden – or rather, the statistical composite of *all* Edens – was not a land of plenty, with fruit hanging in abundance from every tree and fattened calves meekly waiting to be slaughtered. Finite resources – of prey, edible vegetation, water and shelter – could mean that there was competition for those resources. Not just competition between species, but competition *within* the species. In other words, we might well have been at war with one another.

Underlying these material facts are some fundamental truths about gene-centred evolution. In the past fifty or so years we have filled in many of the details about how natural selection works, and its logic yields extremely powerful explanations. At the core of the concept is the idea of the 'selfish' gene. As we shall see, the selfish gene exerts a massive influence on both the evolutionary development and the existing psychology of the human mind. Not only does it affect our struggle for resources and for mates, it also defines the terms on which our sex lives and family lives evolve.

Most of the assumptions we make rest on evidence that is fragile and inconclusive: deposits of animal and hominid bones and teeth, the remains of stone tools, and the patterns of clustering that show hominid camps, hunting grounds or food-processing areas. It remains to be seen whether we are justified in these predictions of group size and the specifics of eating habits. But we should not underestimate the power of archaeology, because old bones can be extremely rich sources of information.

We are learning more and more about the world in which our human ancestors spent their formative years. It was here that our mind and our instincts evolved, and the conditions on the savannah played a vital role in determining which mental adaptations were allowed through the net of natural selection.

Natural selection and adaptation

Physical adaptations are relatively easy to spot. The tongue of the bumblebee is perfectly suited to collecting nectar from inside the deep corolla of a flower. Those species of bee with a short tongue, sometimes called 'nectar robbers', get around the problem by biting a hole in a petal near the base of the corolla and accessing the nectar that way. Similarly, the wings of most birds are well adapted for the purposes of flapping, swooping and soaring. That is not to say that wings were 'designed' to be useful for flying, simply that those birds who inherited genes that coded for more efficient wings were more likely to survive and reproduce. So, in saying that a certain trait 'X' is adapted for a certain function 'Y', I do not mean to impart any sense of intention or design.

The tongue of a bumblebee or a bird's wings are each adept at carrying out their allotted tasks. The elegance, symmetry and efficiency of biological adaptations can be seen everywhere in nature. But evolution is not perfect. We should not fall into the trap of thinking that natural selection turns out the best, 'cheapest' and most elegant solution to any given problem. Many adaptations appear to be the work of a talented and ingenious biological engineer, but there are also examples that seem rough and ready, badly thought-out, or something of a botched job. Our own eyes are one quite good example. True, they have excellent clarity of vision and colour definition. If they are in prime working order they have a fast auto-focus and accurate autoexposure. Additionally, they are self-cleaning and cleverly built into a protective hollow. But we are, many of us, short-sighted,* and cataracts are common. And there is a major

* But there is some suggestion that humans are short-sighted in part, at least, as a result of the demands made by being encouraged to read as children. Given that printing has only existed for around eighteen generations, I guess evolution could not have made much impact.

'design' flaw: the light-sensitive retina lies behind a layer of blood vessels and nerves, and these 'service pipes' limit the amount of light reaching the retina. This arrangement also necessitates a hole in the retina through which the vessels and nerves can pass to connect with the brain – this hole is our blind spot. And, more seriously, it means the retina can become detached rather easily. It would be much better to have the retina in front, and we can find this superior 'design' in large cephalopods such as the squid and octopus.

Evolution is imperfect because it always involves change. The human eye evolved from a patch of light-sensitive cells on the surface of the skin. Once these light-sensitive cells are connected by nerves to the brain and the structural basics of the eye are in place, evolution cannot retrace its steps and completely reorganize the system. The regressive steps would almost certainly never happen.

An added complication is that natural selection encompasses a whole range of simultaneous evolutionary pressures. There is no simple and easy selection for, say, better eyesight without a knock-on effect on another aspect of our person. It may be that the part of the brain used for processing vision uses up power that could have been adapted for some other purpose. No adaptation occurs in isolation, and we are brim-full of compromises and make-do-and-mend.

A good example of this is the malleus and incus, the 'hammer' and 'anvil', the tiny inner-ear bones which in humans transmit and amplify sound vibrations on their way to the ear drum. These minuscule bones have much in common with the bones found in the lower jaws of reptiles, and in the gills of fish. The similarities are present because mammals, modern reptiles and, indeed, fish inherited these bones from a common ancestor. The difference is that during evolution mammals co-opted the bones for their own purposes. The adaptation is therefore limited by the fact that these bones were once used for something completely different; and in order to get from jaw-bone to ear-bone, it could only take steps that produced a selective advantage for the species concerned. Natural selection cannot simply start again from the ground up and choose the best possible solution. If we were to design a human ear from scratch it would be better for not being a modification of something now obsolete. The passenger flying machine a Boeing 747 would have been much less efficient had it been derived from a hot-air balloon.

But for the moment I should simply emphasize that although the human brain may *appear* to be adapted to perform certain tasks, it does not mean that we are looking at *actual* adaptations. In animals, especially those whose behaviour seems to consist completely of instinctual responses and processes, deciding whether or not something is an adaptation is a little easier. Both celestial navigation in birds and the labour performed by worker castes of social insects appear to be the results of natural selection, and are thus embedded in the genes. They are both good solutions to extremely complex problems faced during their evolutionary history. The chances of either of these organisms learning their craft from scratch, as we do with reading, are fairly slim. Equally unlikely is that they are the results of so-called 'genetic drift', which is essentially a series of random mutations.

The human mind is both more complex and more flexible. We are not slaves to our genes, but we are deeply affected by them. Sorting the adaptations from everything else is extremely difficult. We have to try to filter out the signals of the savannah from the chaotic whirl of human activity. Just because certain types of human behaviour are constant across many different cultures does not mean they are genetically determined. As Daniel Dennett points out, all societies who use spears throw them pointy-end-first, but that does not mean we have a species-wide 'pointy-end-first' gene.

Importance of socialization

When babies open their eyes and begin to register the existence of the outside world, there begins an intricate process of neural development. Instincts get switched on one by one, an array of survival tools that come pre-packaged with a newborn baby.

But the brain cannot develop unless it receives the right stimuli.

Susan Greenfield, in her book *The Human Brain*, describes the case of a six-year-old boy who grew up blind in one eye. Ophthalmologists examined all possible causes of his blindness and could find nothing physically wrong. Then, after reviewing his medical history, it was remembered that as a baby he had had his eye bandaged for two weeks to allow a minor infection to heal. As a result, neural circuits that should have processed the incoming signals from this eye had not developed properly – in fact, they were probably co-opted for another purpose entirely – and therefore the eye was useless.

The environment in which we are brought up is critical for the development of human instinct. This is evident in cases where children have grown up deprived of any human presence. The touching film *The Enigma of Kaspar Hauser*, made by Werner Herzog in 1975, is based on a true eighteenth-century story of a young man who spent the first years of his life confined in a cell with no human contact. Once he is mysteriously released, to be looked after by the benign villagers who find him, he is discovered to have lost a number of basic instincts such as speech and fear, and he then has to learn them imperfectly to become more human.

'Feral' children – children who have grown up in the wild, often nurtured by wolves and other wild animals – provide a cryptic glimpse into human nature. There are few genuine examples of feral children, and none of them has been studied with any rigour or objectivity, but nonetheless they provide us with some limited insight into how the development of human nature depends on the people and culture that surround us.

In 1920, the Reverend J. Singh, the founder of a rural orphanage in India, was told of a 'Manush-Bagha', or 'man-ghost', in the jungle some miles from his village. This apparition was said to have the body of a human being and the hideous head of a ghost. The reverend's curiosity was aroused, and he embarked on a trip into the jungle, making sure to take with him a number of armed guards.

They arrived at a huge white ant-mound, as high as a two-storey building. Around it were seven large holes that led to an opening in the centre of the mound. Revd Singh and his party staked out the ant-mound and patiently waited for the arrival of the man-ghost. As dusk was falling, the head of an adult wolf appeared at the opening of one of the holes. It was followed by more wolves and two cubs. Then, crawling after the cubs on all fours, appeared the man-ghost. The body was that of a human child, truly a kind of caveman, the head a big matted ball staring out of which, Singh could see, were piercing eyes.

The man-ghost ran off, still on all fours, into the jungle. Singh decided to come back with men and tools in order to demolish the ant mound and smoke out its inhabitants. As the first spadefuls of earth were carved out of the mound, one of the wolves, an adult female, ran out, and she was shot by one of Singh's men. As they dug further into the core of the ant-mound they found two cubs and two 'ghosts' huddled in a corner who struggled with their captors, baring their teeth, but eventually the men bundled them into sheets and hauled them away.

The ghosts were two young girls, one an infant, their faces almost completely hidden by a wild mass of matted hair. Eventually they became calmer, and Singh was able to feed them raw milk and water. He cut the two girls' enormous mass of hair, and named them Kamala and Amala. Kamala, he guessed, was around eight years old, and Amala was about eighteen months.

Behaviourally, the two girls were not human, that much was clear. They bore the marks of their life in the wild, and were covered in sores, cuts and boils. Their joints had seized up so they could only move about on all fours; they certainly could not stand up. They appeared to be nocturnal, sleepy during the day and lively and awake at night. Singh reports that their vision appeared to have become adapted to the night, and they had an ability to see in the dark with ease, although on this point he stretches our credulity. They had a taste for raw meat, too. Kamala could smell meat from some distance, and once she was caught eating the entrails of a fowl that had been thrown outside the orphanage compound. They did not use their hands to eat, but instead lowered their mouths to the plate or bowl, like a wolf would do. They showed no signs of communication, and it would be some time before Singh heard them utter a single meaningful sound. They would urinate or defecate anywhere and at any time, and slept entwined together like kittens.

It was not long before the girls, seemingly in need of protection and affection, started to show an attachment towards Singh's wife. They steadily became more trusting and more playful, but there was still no sign of language, or comprehension of gestures beyond the most basic level.

Less than a year had passed when both girls became seriously ill with dysentery. They were found to be infested with roundworms. They became weaker, and only moved when drink or medicine was brought to their lips. Then, Kamala, the elder of the two girls, showed signs of recovery, but the very next day Amala died. Singh reports that when Amala died, Kamala refused to leave the body. She touched Amala's face, tried to open her eyelids. Singh says that two tears dropped from her eyes. For the next six days Kamala sat in a corner, alone, unresponsive to affection or touch.

Kamala would sometimes overeat, so much so that she made herself ill. She would not kill animals herself, but if she found carrion she would carry it home, sometimes driving away vultures from the carcass. Occasionally she would hide the carcasses around the orphanage. She also developed a taste for sweets.

Slowly, she began to pick up a few words, like 'yes' and the word for 'dress', and she knew the names of some of the babies in the orphanage. She understood colours, and became sufficiently acculturated to use the lavatory, at least whenever she thought the Singhs were watching. But she did not progress beyond this. After nine years of living with the Singhs and their foundlings, Kamala fell ill once again, and died.

The wolf-girls of Midnapore could never express themselves beyond communicating the most basic needs. Their story makes clear the fundamental importance of nurture, experience and the social environment in which we are brought up. Our cognitive mechanisms for dealing with the world – whether they are face recognition, language acquisition or emotional development – will not appear of their own accord. Beyond a certain point, it may be too late to 'switch' them on.

Just like the development of a child, the process of evolution is intertwined with the growth of culture, and 'culture' began well before evolution finally shaped us as we are today. Professor Geertz reminds us that although the invention of the aeroplane has not produced any biological adaptations in humans, that is not necessarily the case for tools invented a couple of million years ago. Stone tools are part of hominid culture, and the simple chopper or flake, invented by *Homo habilis,* which was made by smashing two rocks together, might have affected the evolution of our opposing thumb, our posture, the size of our teeth and, most importantly, mental capacities like dexterity or spatial reasoning.

It's tempting to look at our modern lives – our drives, our desires, our hopes, our troubles – and find a nice, simple evolutionary explanation for them all. But we need to be careful. Clifford Geertz proposes that the study of mankind should not entail the simplification of complex explanations. Instead, he says, it should comprise the substitution of simple pictures with complex ones while striving to retain the clarity that went with the simple ones.

Our evolutionary past exerts the most powerful pressure. But the genetic element of human behaviour will always be refracted through the medium of culture. Genes are responsible for the human mind in the same way as a scriptwriter is responsible for a movie. The script forms the basis of the film, but the look and style of the film is determined by the director, designer, editor, and so on. Some of the dialogue will be improvised on set, and occasionally the writer gets fired and the script gets rewritten. And, as my favourite playwright, Luigi Pirandello, might well have observed, each person watching the film may have a different interpretation.

This book is an exploration of the theories and discoveries that populate what is a relatively new area of evolutionary theory. I should perhaps add that it has been written to accompany the television series called *Human Instinct* and, as I mention in my acknowledgements, is heavily influenced by friends at the BBC who made those programmes and by other, equally talented friends making another series for television, *Walking with Cavemen*, with which it has been my privilege to be involved.

Nearly all of the best thoughts in this book are, sadly, not mine, but I take full responsibility for any mistakes, failures of interpretation and assumptions. Moreover, many of the most interesting theories are contentious, and some have long provoked outright hostility. But I do hope that this book may shed some small light on what it means to be human, on what it means to be a product of evolution not just physically, but mentally. Instinct, an invisible hand, is ever present in all our lives, and revealing its emotional form may just allow us to understand our real selves a little better.

chapter**one**

The Origins of Survival

Fight or flight

You are walking home late one dark, wet and misty winter evening. It's been tiring today, so you are keen to get inside, close the door behind you and put the stresses of the day to rest. As you amble along, thinking in neutral, you realize you can hear measured but quickening footsteps behind you. A snatched glance over your shoulder reveals a man approaching in the gloomy street-light. He is moving rather quicker than you are walking, and he is looking at you all the while. There's no-one else in sight – no-one on the street except you and the stranger. The house suddenly seems a long way away. In less than an instant, you suddenly feel very afraid. Your heart starts beating wildly, your mouth goes very dry and you have a huge urge to start running towards the safety beyond your front door.

There is a very simple reason why you feel so terrified. Inside your body, all hell has broken loose. Biological sirens and alarms are wailing. Perceiving the threat of the potential mugger with lightning speed, your brain and autonomic nervous system – the automatic controller of the gut, heart, vessels and lungs – have gone into overdrive and produced a huge surge of adrenalin. This triggers a hormonal cascade inside you, an incredibly fast and powerful chemical relay-race designed to propel you away from a threatening situation. Just a fraction of a second later, the hypothalamus in your brain begins pumping out a substance called corticotropin-releasing hormone, or CRH, which in turn sends alarms to the pituitary gland in your brain to pour out adrenocorticotropin, or ACTH. Finally, the abnormally high levels of ACTH in your bloodstream are the warning signal for the adrenal glands, near your kidneys, to start producing cortisol.

Imagine the speed at which these precise yet complex combinations of hormones are produced – your body's reaction to fright and attack is virtually instantaneous. Almost immediately these chemical alarm bells are set off inside us, we are forcibly shoved into the (all too familiar) feeling of being acutely afraid. The adrenalin makes your heart pound faster, increasing its normal resting rate by as much as two or three times. You would have to cycle really vigorously for maybe fifteen minutes to produce that kind of rise in your heart rate under normal circumstances, but in the sudden grip of fear, the rate can triple in just a matter of seconds. You are also breathing much faster now and the blood is being rapidly redistributed around your body. The blood vessels in unimportant areas like your stomach and your skin constrict, shunting blood away and into the now dilated vessels of the muscles of the limbs. Here the extra oxygen and fuel gained by your increased breathing can be best harnessed to flee from the threat, or even fight it. There wouldn't be much sense in your stomach busily digesting that lunchtime sandwich right now, when every drop of your body's available energy needs to be used to save you from the approaching threat.

As the adrenalin and cortisol continue to gush out into your blood, your pupils dilate, allowing you to see better in darkness and shadows and to perceive any movement around you more keenly. A kind of pain-dampening effect is switched on so that you won't be distracted from getting away by any injuries. Emergency reserves of glucose are released inside you to allow for especially intense bursts of muscular activity. Even your immune system is mobilizing to cope with the possibility of dealing with a serious wound. In just a matter of moments, your body has propelled you into a state of extreme physical and psychological readiness to run or fight – whichever course of action best suits the threatening situation.

As the stranger, now just feet away from you, holds out the single, familiar glove you now realize you'd obviously dropped some way back, you may ask whether all your body's efforts were really worth it. Whether it's 'butterflies' in the tummy before an interview, the dry mouth and throat we feel minutes before we have to make a speech, or even the quickening pulse and sudden jump as we hear an unexpected bump in the night, it often seems that our bodies are overreacting. So where does this physical and psychological reaction come from? It's not as if we were taught as children to start breathing faster in threatening situations, nor can we consciously make our heart beat so much faster or force our body to produce adrenalin. What we are actually experiencing is our very own personal link to our most ancient human ancestors – a reaction which hundreds of thousands of years ago almost certainly made the difference between life and death, but which now, in most cases, simply serves to remind us of the remarkable fact that while living in a very advanced modern world, we all do so with Stone Age brains and bodies.

Indeed, this reaction in response to stress hormones goes way back in time, well before our immediate ancestors. Even animals that aren't mammals react basically in a similar way. Try startling your goldfish as it meanders around its bowl. If you place a net or a threatening object into its water, you will immediately see a very similar kind of reaction. Its fins stand out ready to flee and the gills and mouth start opening and closing in overtime. That fright reaction is caused by the same hormone, inherited down the ages: adrenalin.

Our early human ancestors lived in a very dangerous and threatening environment. When they first made it out of the trees to try their luck on the grassy plains of the east African savannah, they were vastly outnumbered by vicious and hungry predators. They lacked the brute strength of the great apes and many other large land mammals, especially the big cats. Nor were they particularly fast or agile like the antelope or gazelles. They could not fly, nor were they especially well designed for life in water. Their senses were poorly developed: no night vision, no extra-sensitive hearing to detect prey rustling in the grass hundreds of feet away, and an extremely unsophisticated sense of smell. Ape-man infants were helpless and dependent, and parents were distracted from practical matters of survival by having to care for their young. But these naked and defenceless prototype humans had to contend with the searing African heat as they traversed the vast distances of the plains in search of food, shelter and mates. If they stayed in one area, they risked starvation and attack from a stalking predator; if they were on the move, they faced the test of the unknown, of coming face to face with some terrible beast. And terrible they were.

While the most probable threat you may experience today is a brush with a suspicious person in a local street, our ancient ancestors had to face the reality of encounters with violent sabre-toothed cats and other predators. One variety was Smilodon, a sabre-toothed cat whose remains show it was almost a foot shorter than a modern-day lion, but weighed almost twice as much. Instead of the long, graceful tail cheetahs and leopards use for balance as they race across the African plains, Smilodon had a short, stumpy bobtail. This beast was a simmering hunk of muscle, designed for quick and furious violence.

Smilodon almost certainly hunted in packs. We know this because fossil specimens of the huge cat, dug up in California, show evidence of healing injuries. Some of these injuries were so serious that immediately after the trauma the cat would have been unable to hunt, so it couldn't have survived long enough for the injury to heal unless other animals from the pack had brought it food. Smilodon could roar – we know that from the structure of the hyoid bones in its throat – but like any modern feline predator it would have been silent when stalking. Once it had ambushed its prey, by charging the frightened antelope or artiodactyl with an explosive burst of power, it would have used its long, curved, sabre-like teeth, viciously ripping open the belly or throat.

It's highly likely that predators such as these would have killed and eaten early humans. In a cave at Swartkrans in South Africa, palaeontologists found the skull of an early human, *Homo habilis*, buried deep in sediment dated to around two million years. It belonged to an eleven-year-old child and bears the mark of an

African predator: the bone is punctured in two places, an exact match for a pair of leopard's canines. A child such as this stood absolutely no chance of defending itself against these powerful beasts; even a fully grown male would have been practically helpless, given the speed, power and aggression of the big cats.

To stand the best possible chance of survival all animals have to protect themselves from danger and death, so they need a means to be alert to threats at all times, to fear them and to fight or flee in response to them. The imperative is self-preservation as well as the survival of the species. In evolutionary terms, a fearless animal would be much less likely to survive and pass on its fearless genes. Six billion humans now populate the world; our species has become the most successful in the history of all life on the planet. Our early ancestors must have developed and evolved some spectacularly successful ways to protect themselves from predators and threats – physiological and psychological reactions that were so fundamental to their survival they still exist deep within us today.

The seat of instinct

The control centre for all our instincts is our brain and spinal cord. In the 1950s, Dr Paul MacLean, the distinguished neuropsychologist from the National Institutes of Mental Health in the USA, proposed the idea that the brain could be regarded as being in three parts. He called his theory 'the triune brain'. He believed that as we evolved from amphibians into land-living mammals and then into primates, our brains were enlarged not by a complete restructuring or reorganization, but by building 'extensions' onto the ancient inner core – more advanced improvements, if you like.

First came the so-called 'reptilian' brain, the ancient inner core present in all reptiles which supports the most basic functions of breathing, blood circulation and digestion. It is also involved in some pretty basic aspects of behaviour: mating, aggression and anger. In humans, this reptilian brain sits above the spinal cord at the base of the rest of the brain.

Wrapped around this basic brain is what MacLean called the

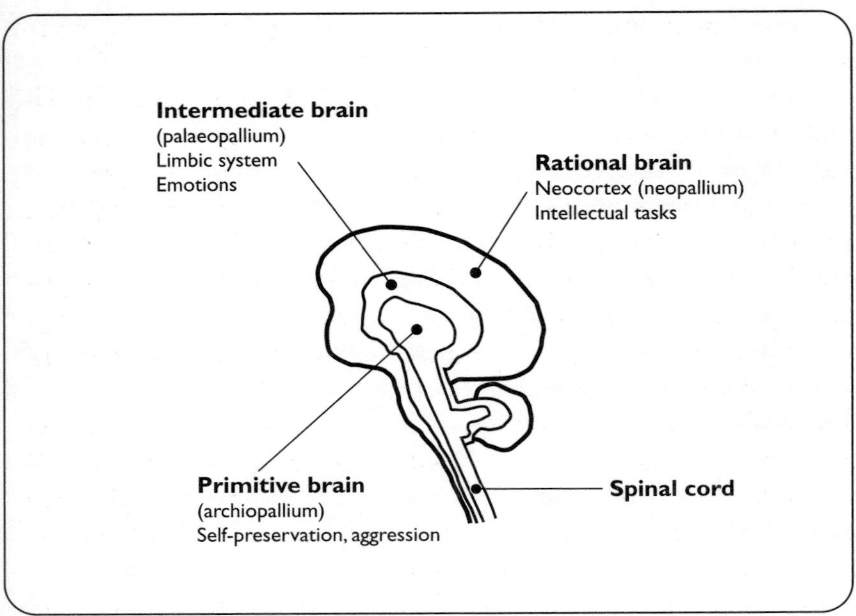

THE TRIUNE BRAIN

'limbic brain'. Without a limbic brain, you're only going to have the most basic repertoire of emotions. The limbic-less reptiles don't even care about the fate of their offspring, usually abandoning their eggs or occasionally even tucking into them for dinner. When we were filming the television series that accompanies this book, I got a chance to see this loveless type of parenting at first hand. We watched as a leatherback turtle, a massive Central American sea creature, laid her eggs and loosely buried them in the sand, only to leave the tiny hatchling turtles entirely on their own to negotiate a hazardous route to the sea. Most are destroyed by predators. It wasn't until evolution proceeded and higher animals developed a limbic brain that many of the more basic emotions developed, including the protection of the young. Indeed, feelings such as love, sadness and jealousy appear to have their roots in the limbic brain.

Evolution of a still larger brain provided the third major component, the neocortex. In MacLean's model, over time, the neocortical brain provided logic and thought and in humans led to such processes as speaking, planning and writing.

Although there are now a number of legitimate objections to Dr MacLean's ideas about subdividing the brain so rigidly, his general approach is a good reflection of our understanding of how the mammalian brain evolved. It is largely true that the limbic brain can be thought of as the area from which many, if not all, of our instincts arise. It comprises areas such as the hippocampus, thalamus, hypothalamus and amygdala which are involved in memory and much of the behaviour related to sex, hormones, food, the perception of pleasure and competition with others.

Nowadays, our understanding of the neurology of human instinct has reached an unprecedented level. Recent advances in our knowledge of the brain and how it functions have thrown new light onto many of its secrets. One of the most fascinating areas has been the study of the very seat of the fight-or-flight response – the neurological home of fear. It measures just an inch long, resides deep inside your brain and is shaped a bit like an almond. It's called the amygdala.

Pinpointing fear

For most of the history of medicine, the brain yielded few of its incredible secrets. Pickled in alcohol or formalin after death, the mass of grey matter is bland and relatively shapeless, with the constituency of a rubbery old mushroom. How could this heavy lump of an organ tell doctors anything about the remarkably complex actions, thoughts and feelings of its now deceased owner? Dissection proved to be pretty useless. There were no visual clues to suggest where different functions and abilities were located, and the tissues of the brain are of uniform density, so X-rays simply produced murky shadows. Neurologists and pathologists could piece together some information from examining the brains of people who, when they were alive, had some particular physical or mental disability. Sudden inability to see colour or a severe form of memory loss, for instance, may correspond with an area of the brain which has been physically damaged by injury or a stroke – but more often than not, a damaged or abnormal brain appears to be completely unremarkable. What chance did neurologists have of working out exactly where in the brain basic functions like breathing and sleeping were controlled? How could they locate the residence of much more complex abilities such as memory? As for the intricate emotions of loving, hating, being afraid, being disgusted – there seemed little chance.

But an enormous advance, the advent of a technique called magnetic resonance imaging, or MRI, has revolutionized our understanding of the workings of the living, functioning brain. We're now able to construct an actual image of how the brain thinks and feels. By taking multiple scans of the brain and analysing the results on computers, we're able to build up a detailed three-dimensional picture.

How does MRI work? We don't always think of our bodies in this way, but the truth is we're not mere flesh and blood. Each part of the body is also made up of atoms, which are rather like miniature magnets, each with a north and south pole, just like the Earth. If a part of the human body is put inside a scanning machine which contains an intense magnetic field, the atoms in the tissues will line up like magnets, with their north poles all facing in the same direction. Now, with a quick blast of radio-waves at the lined-up atoms, they would all start to rotate. As they return to their north–south orientation, the atoms themselves would emit radio-waves and these would be picked up by sensors in the machine. But the clever trick is that in every different tissue of the body, the atoms happen to rotate at different speeds, so each tissue can be singled out on the scanner and added to the overall picture to build up the final image.

What I find remarkable is that, although the principle behind MRI was first recognized in 1946 – winning its discoverers, Felix Bloch and Edward Purcell, the Nobel Prize – it was never initially considered suitable for medical use. People thought it only useful for looking at inert substances and man-made structures. It wasn't until 1971 that scientists began to wonder whether MRI might be used to look at living biological tissues, and there was huge argument over whether the atoms in the body would resonate and whether any kind of image would be remotely possible.

In the early 1970s, I was working at Hammersmith Hospital looking at how eggs were transported inside the Fallopian tubes of rabbits. This was an important study because it had implications for new methods of contraception which were potentially useful, particularly in the developing world where the population explosion was producing such social pressures. At that time, coincidentally, Hammersmith was the first medical establishment to invest in researching and then using MRI. Scanners nowadays cost around a million pounds; my colleagues doing experimental work built their own. They used bits of wire, transistors, obsolescent radio components, primitive computing equipment, scrap metal and even adhesive tape and string. These bearded, nerd-like acquaintances who seldom emerged from the deep, darkened sub-basement (except to drink) persuaded me after a session drinking dangerous half-pints of lager in the bar that I might be able to see the innards of one of my rabbits. I would need to descend into subterranean parts of Hammersmith that few people working on that site even knew existed. They assured me their machine wouldn't hurt my rabbit (though I found out much later that this was the first bunny they had tried to look at). Nonetheless, I felt deeply sorry for Laura as I put her into this bizarre, humming machine with its huge electromagnets on the outside. But we all jumped up and down with excitement when after about fifty minutes of scanning we saw a faint image of the contents of my rabbit's abdomen on a primitive cathode ray screen which I think had been salvaged from the Army & Navy surplus stores. Laura didn't seem particularly impressed – perhaps because the picture was mostly like a blurry snowstorm. Try as I might, I realized I couldn't work out the anatomy of my rabbits from the pictures I was getting, so I abandoned the experiments.

But only ten years later, long after I had lost personal interest, MRI was to revolutionize medical imaging. After years of developing the technique to image the whole body in the 1970s, it had been honed into a tool used particularly for scanning the brain, as it was realized that the different tissues of the brain give off a slightly different signal, thus producing excellent and very detailed MRI images. In 1992, the technique evolved even further, into functional

MRI (fMRI), which allows us to take an extraordinary peep into the actual functioning of the different regions of the brain. We were able to see minute changes deep within the brain as they occurred, we could test intelligence, and soon we could ask people questions as they lay in the scanner and then watch as the different areas of the brain lit up in response. Now, functional MRI allows us to build up a picture which pinpoints the precise areas of the brain where actions are controlled, where thoughts take place and even where emotions are felt – and to watch as they happen.

This advance was made at the time of another very important new technology: positron emission tomography (PET). The use of this second brain-imaging technique – during which we inject a tiny amount of a very short-lived radio-isotope, or radioactive substance, into the body and watch as the particles the isotope emits collide with the body's electrons – has given beautiful computed pictures of the brain. Using PET, we can measure blood flow, for instance, and changing metabolism within the brain.

Initial uses of functional imaging, both fMRI and PET scanning, were explorations of the response to sensory stimuli. Early researchers were delighted to see the visual cortex at the back of the brain glow brightly when subjects in a scanner were exposed to flashes of light. More recently, my colleagues at Hammersmith Hospital have used PET scanning to evaluate painful stimuli (in some most altruistic volunteers). But these techniques have developed rapidly and are now much more sophisticated, as we shall see. For the very first time in human history we know some of the remarkable truths about what it is that 'makes' you feel excited, depressed, hurt or afraid, and why.

The amygdala

A few years ago, a woman – who for the sake of anonymity is referred to in medical casebooks as SM – turned up at her local hospital in Iowa, USA, suffering from epilepsy. When doctors began to look for the root of her condition using an MRI scanner, they made an incredible discovery: her entire amygdala on both sides of the brain had been destroyed. In fact, it turned out that she suffered from an extremely rare condition called Urbach-Wiethe disease in which calcium deposits are laid down in the amygdala over time, eventually destroying it.

SM's case was a unique opportunity for researchers to uncover the pivotal role of this tiny almond-shaped part of the brain in processing emotions. They began subjecting her to a battery of psychological tests, showing her photographs of people expressing various emotions and assessing whether she could understand or even register them. She categorically failed what the researchers called 'the Doris Day test'. They showed her a film clip of the actress Doris Day screaming with fear. SM was baffled, asking doctors 'What on earth is she doing?' In fact, she was utterly confused by any picture showing someone who looked afraid. She also had problems deciphering various other negative emotions, such as anger and surprise. In complete contrast, she had a totally normal response to positive emotions such as happiness and joy.

It's certainly a reflection of just how vital the fear reaction really was for the survival of our ancient ancestors that a special, dedicated brain system evolved to deal with it. The location of the amygdala gives us a clue to the precise nature of its role in processing fear. It has connections to the autonomic nervous system – which, as you'll remember, controls physiological reflexes such as your heart and breathing rates – as well as to other brain regions that process sensory input. It's like a neurological crossroads, the hub of a network of pathways in the brain, a special 'rapid response unit', if you like, primed to act quickly when presented with danger.

Here's how it works. You're walking through the woods and out of the corner of your eye you spot a long, smooth, curved object on the path ahead. Before you can even think 'Snake!' your amygdala has triggered the fear response and the biochemical cascade inside you has begun, preparing your body for this possible threat. The amygdala recognizes snakes as representing danger, so the body's physical resources need to be at the ready – to fight or flee.

But what's interesting about the amygdala is that it reveals how even we big-brained humans are actually designed to act first and think later. The work of neuroscientist Joseph LeDoux, a professor

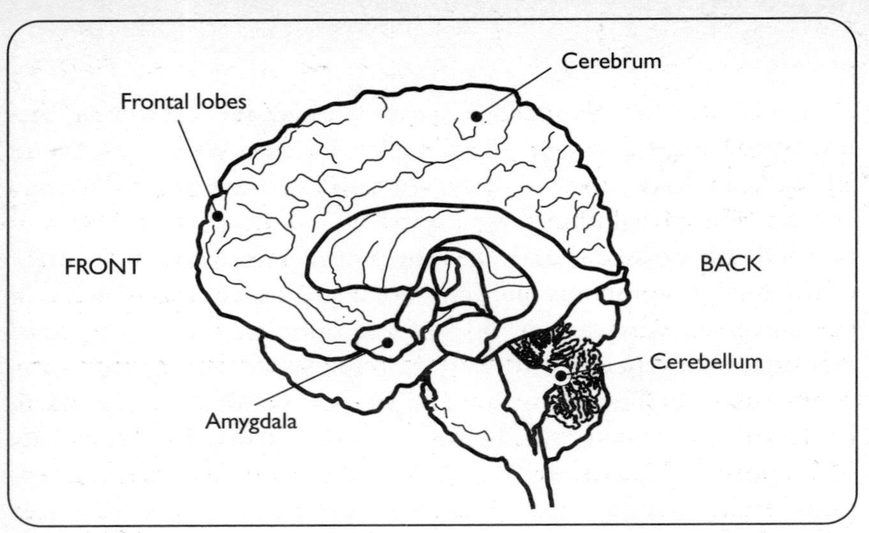

THE AMYGDALA

FROM THE SIDE

FRONT

FROM ABOVE

BACK

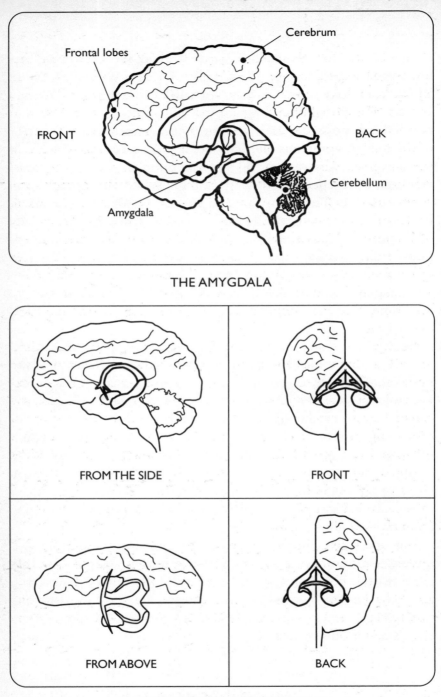

THE STRUCTURE OF THE BRAIN

at the Center for Neural Science at New York University, has uncovered some fascinating facts about our fear response. LeDoux and his team have shown that the amygdala's connection to the eyes and ears is a special ultra-fast pathway which means it has access to raw and unprocessed sensory information. There is no involvement of the higher, conscious brain, no cognitive processing which tells the amygdala what to do. So it turns out that our most primal emotion, to advance or retreat, is triggered so fast that it precedes all conscious thought and awareness. In fact, as Dr David Amaral, Professor of Psychiatry in the Center for Neuroscience, University of California at Davis, has discovered, there are far more connections leading from the amygdala to the prefrontal cortex – the area of the brain most responsible for planning and reasoning – than there are going in the other direction, which may be one reason why we sometimes find it so difficult to exert conscious control and logic over our fear.

Because our brains are wired to feel before they think and indeed to feel at lightning-fast speed, our amygdala will often make mistakes. That object on the path may not actually be a snake. Perhaps that state of high alert isn't needed after all. Eventually, the cerebral cortex catches up with the amygdala and applies its superior processing power to the problem. The object on the path isn't moving, it has no head or tail and bears none of the markings normally found on the skin of a snake. In fact, it's just a bit of twig. Conscious thought has intervened, stepping in to allow our muscles to relax, our heart rate to slow down and the adrenalin pumping around our body to subside.

But don't give the amygdala a hard time for making the odd mistake. Emotion paves the road for conscious thought for one very good reason: if it didn't, we'd all have been bitten by snakes long ago. The bottom line is that in a threatening situation, thinking simply takes far too long, and when our survival's at stake we are always better off safe than dead.

The evolutionary roots of fear

There was a very clear advantage for our ancestors to feel fear: it enabled them to switch on their defence mechanisms, steady their limbs for action and perform incredible feats of survival. Indeed, modern-day experiences reveal the extraordinary powers of adrenalin; we've all heard the occasional tale of someone who's fought a fifteen-foot-long python or knocked out a fully grown black bear with one left jab. There are even instances when the instinctive reaction of fear appears to have defied the limits of human endurance. Could a grandmother really lift up a pick-up truck to snatch out a trapped child from under a wheel?

The immediate unconscious response that triggers the powerful biochemical cascade is clearly a classic example of human instinct. But where does all this come from? The fact that the process is fast and automatic strongly suggests that we have a programme hard-wired into our brain at birth which determines how we react to danger. The fact that all other mammals share this particular mechanism also suggests that we are born with this instinct. But let's take a closer look. How do we know what to react to? Are we really pre-programmed to react to snakes and other dangerous creatures with fear and loathing? Is every baby born with some kind of genetic memory of a snake etched into its little brain so that it will freeze at the first sight of a snake, or even a curved stick, on a forest path?

Charles Darwin certainly believed that this was the case. 'May we not suspect', he wrote, 'that the vague but very real fears of children, which are quite independent of experience, are inherited effects of real dangers and abject superstitions during ancient savage times?' He went on to reflect that these childhood fears often disappear later in life. Although it's normal for a child to be afraid of the dark, the fear is not usually continued into adulthood. Darwin thought that such fears might be similar to anatomical structures which seem to be genetic 'throwbacks'. At a certain stage in its development the human fetus displays paddle-like limbs, and during the early growth of the embryo there are even gill-like structures. The paddles are, of course, soon replaced by hands and feet and the 'gills' rapidly disappear completely, unless the developing human grows abnormally. Such structures can be thought of as remnants of an earlier stage in the history of human evolution. So perhaps there really are primitive mental remnants that work in much the same way, to a greater or lesser degree from person to person.

But was Darwin right? Can our fears really be inborn and hard-wired when we also know just how important the effects of learning are as we grow up? Could it be that it's the experiences we gain from our environment which, among so many other lessons, teach us what we should and shouldn't be afraid of?

Pavlov and the malleability of the mind

When I was a schoolboy of twelve, Ivan Petrovich Pavlov was one of my first scientific heroes. I vividly remember how a copy of his translated *Conditioned Reflexes* sneaked from my father's bookshelves thrilled me with the fascination of the study of physiology and the workings of the body. Pavlov was born in 1849 in Ryazan, a small village in central Russia. Although his family wanted him to become a priest, he was influenced by the eminent Russian physiologists of the day and by Charles Darwin's writings and determined to pursue a career in science. Although initially he was most interested in the mechanics of digestion and how the digestive juices worked, his experiments with dogs paved the way for understanding the nature of 'conditioned' or learned reflexes.

These pioneering experiments showed that some very basic types of behaviour, similar to instincts which influence the way the organs in our body function, could be learned. Pavlov's most famous experiment involved showing a group of dogs a bright light a few seconds before their food arrived. Again and again the pattern was repeated; the light went on, and after a matter of seconds the food would appear. After several days of this repetition the dogs began to salivate as soon as the light was turned on, even if the food was nowhere to be seen (or smelled). These malleable beasts had been programmed successfully; their brains now anticipated the delivery of a meal, just as if the aroma of the food had found its way into their nostrils, and their bodies began to produce saliva in preparation for digesting it.

Pavlov tried a variation. He rang a bell before the food appeared – and exactly the same thing happened. The dogs' most fundamental mental processes did appear to be responsive to learning in a way no-one had previously thought possible. But what was also of interest was that if the bells were repeatedly rung and no food ever appeared, Pavlov's dogs eventually stopped salivating when they heard the bell sound.

Pavlov himself had rather little time for psychiatry and was more interested in the workings of the body than in the workings of the mind. But he did wonder whether his observations of conditioned reflexes in dogs might go some way towards explaining the behaviour of psychotic people. His dogs became the adopted mascots of the 'behaviourists' – a breed of psychologists who believed the mind was infinitely malleable, moulded by learning and our ongoing life experiences of pain, pleasure, fear and happiness. They constructed a complex set of theories with the human mind as a 'tabula rasa', or blank slate, ready to be programmed by the events we experience from birth onwards.

If you support the behaviourists' way of thinking, you may put your brain's reaction to a snake down to seeing your older brother being terrified by one when you were a young child, or to having a nasty brush with one yourself. That lifelong anti-snake response might even have been triggered by watching the scene in *Raiders of the Lost Ark* when Harrison Ford gets thrown into the horrible snake pit.

Fear is one thing, but phobias are quite another. Phobias are commonly defined as persistent and anxiety-inducing fears which are completely out of proportion to the actual danger. I think it's pretty unlikely I'll ever be thrown into a pit full of poisonous snakes, in fact I'll probably only ever see one in a safe environment like a zoo, but my fear impulse may be fired up just the same. If this reaction were to become a repeated overreaction, it would be a phobia.

Pavlov, had he been so inclined, might have been able to make his dogs develop phobias. Instead of giving them food after the bell was rung, he could have administered an electric shock, rather like the aversion therapy practised on the character Alex in Stanley Kubrick's film *A Clockwork Orange*. Every time the bell was rung they would have been terrified, even if a shock never came. Could we 'learn' to have phobias in the same way, through experience? It certainly seems difficult to find any kind of evolutionary explanation for some of the more bizarre phobias that have been described in the psychiatric journals. Take a look at an alphabetical list of known phobias and you'll see what I mean. You'll find arachibutyrophobia, a fear of peanut butter; enetophobia, a fear of pins; then erythrophobia, fear of blushing. There's coulrophobia, a fear of clowns; lutraphobia, the fear of otters; even xanthophobia - fear of the colour yellow. How on earth could we link any of these back to our time on the savannah? I don't think clowns ever played much of a part in our evolutionary past.*

A clever experiment was devised to try to work out to what extent fears were learned from family and others who are around us during our childhood. A group of rhesus monkeys which had been reared in a laboratory were shown some live snakes. These monkeys, who had never come across snakes before, showed absolutely no fear; there was no sign of anxiety at all. At least for this particular species, there appears to be no hard-wired, genetic fear of snakes that manifests itself from birth. But wild monkeys do exhibit a fear of snakes. Could the lab-reared monkeys 'learn' this fear from their wild cousins? They were shown wild monkeys coming face to face with a live boa constrictor and reacting with terror. Interestingly, when the lab-reared monkeys were then put next to a snake, even a toy snake, they showed the same terrified reaction. They had indeed learned how to be scared, and more experiments at a later date showed the effect to be permanent. The experiment worked even when the lab monkeys were shown a video of the wild monkeys' confrontation with the boa constrictor.

So far, so good. Fears appear to be taught and the monkey mind

* But of course, clowns obscure their face with paint, thus covering their features and true emotions; they confuse our ability to judge their mood. In short, they are really quite sinister – a point several film-makers have recognized.

seems to be malleable, just like Pavlov's dogs – but wait for the twist in the tale! Using a magician's deviousness, the researchers set up another experiment so that the wild monkeys were once again confronted by a snake, but owing to the clever use of mirrors, the lab monkeys watching from the sidelines saw a bunch of flowers in place of the snake. The wild monkeys, of course, delivered the same terrified reaction. As with snakes, the lab monkeys had had no prior experience of flowers. They had no idea whether they were dangerous or not. Would the lab monkeys acquire a fear of flowers in the same way, just because they'd watched the wild monkeys being apparently afraid of them? The answer is no. The monkeys showed no anxiety of any kind when they were later confronted with a bunch of geraniums.

It's almost as if the fear of snakes lies dormant in the monkeys' mental hardware, ready to be switched on when the conditions command it or when other monkeys 'show' them the correct reaction. They do not have the ability to be scared of flowers, however hard researchers may try to instil one. So monkeys' fear is driven by an elaborate combination of hard-wired instinct and experience, which can turn on particular fear circuits in their brains.

E. O. Wilson, one of the founders of sociobiology – the study of the evolution of animal social behaviour – compared this process to exposing an image on photographic film. After exposure the image is present but invisible; it appears only when the film is developed. The image may be developed to varying degrees of darkness or lightness, with a different tint or contrast, or perhaps not developed at all, but the picture itself remains the same. Our instinct for fearing snakes has always been imprinted on the film and all that remains is for the film to be developed.

Pavlov did not take evolution into account. The snakes and flowers experiment supports an alternative theory of the primate mind and reveals the powerful role of our evolutionary history in our modern psychological make-up. Instead of the tabula rasa, the blank slate, with its malleability and ability to learn from scratch, it appears as though the learning of phobias takes place with certain clearly defined limits, embedded deep in our genes.

Indeed, putting the rarer forms of human phobias aside, the vast majority of the rest fall into one of four very clear categories, all of which would have indeed had meaning for our ancient ancestors on the savannah: firstly a fear of animals such as snakes, spiders and insects; then a fear of natural environments such as heights or the dark; fear of blood or injury; and finally a fear of dangerous situations such as being trapped in a tight space. For early humans, fearing such threats would clearly have been a useful mechanism in prolonging survival: those who possessed the ability to sense and evade these dangers were more likely to live and reproduce, passing those fears on to future generations. Times have changed and new fears have developed, but the ancient programming of our brains is still clearly evident today.

And in contemporary versions of Pavlov's experiments, some fascinating results are helping to pinpoint where in our genes this kind of fear programming may lie. Two different research studies, set up by the National Institutes of Mental Health in the USA, exposed mice to a light or the sound of a buzzer before giving them a very mild electric shock. The mice, of course, learned to associate the cue with the shock and became paralysed with fear when the stimulus was switched on, even if no shock followed. But interestingly, these studies also showed that some mice had a much stronger fear response than others. Using a variety of complex markers, the researchers were able to search the mouse genome for a gene which contributed to this differing response, and they believe they have located it on chromosome 1.

A recent human genetic study of people suffering from anxiety or panic disorders by Dr Gratacòs and colleagues in Spain has also focused interest on an apparently important stretch of DNA on chromosome 15. These researchers found that in some cases of phobia, the DNA on part of this chromosome may be duplicated or repeated. This particular region of the chromosome is relatively short in genetic terms – about seventeen million letters – and contains some fifty-nine genes. At least four of these genes produce proteins that may be important in brain function – neurotrophin 3 receptor and three variations of the nicotinic acetylcholine receptor. But what is also of interest is that a tiny proportion of people with this chromosomal aberration do not suffer undue anxiety, panic or phobia. This argues in favour of the possible presence of other genes in these people which control the region in this chromosome and prevent these adverse effects.

But, as Dr Jonathan Flint of the Wellcome Trust Centre of Human Genetics in Oxford has said, undertaking genetic linkage in behavioural disorders is not for the faint-hearted; few areas of human genetics have generated more acrimony. 'All the vicissitudes the field has endured would occupy at least one book,' Flint points out. The geneticist has to sort out the real problems when many genes interplay with one another in addition to classifying and delineating a specific psychiatric behaviour. And, as Flint says, psychiatric classification, itself not free from dispute, may be reliable, but it does not always correspond with biology. Even in closely related people such as twins, a genetic predisposition for autism or schizophrenia can give widely differing outcomes in the same family.

What effect these bits of DNA may have on the human brain is not yet clear, but it is very likely that certain types of receptors in the brain process the fear response. Clearly it's important to recognize that we're not talking about a single gene 'for' phobias, but we are potentially standing at an exciting first point in uncovering the fundamental genetic roots of the complex emotion of fear which may just be something inherited to varying degrees by all of us. As we shall see in later chapters, there is increasing support for an evolutionary perspective on the human mind, but there are dangers in pursuing this view to the extreme. I think we need to be careful in presenting human behaviour as simply a pre-defined product of our evolutionary past.

Ancient memories

Remember how the monkeys learned, or 'relearned', their fear of snakes? It was as if they had a kind of 'genetic memory' which we saw was switched on in a particular social environment. Could we, too, have genetic memories handed down to us by our hominid ancestors, memories which were formed over millions of years in the grindingly slow process of natural selection – our very own links to our evolutionary past?

The incredibly slow pace of evolution would certainly have allowed for such 'memories' to be etched into our genes. Let's consider the physical landscape in which *Australopithecus*, *Homo erectus* and the other prototype humans lived. After the forests dried out and we spilled onto the savannah, we spent over three million years living in this one probably constant environment. Think of the huge number of generations who experienced it as their home, who were adapted to it, who passed those adaptations on to future generations. If I were to delve deep into your own mind today, might I find some kind of genetic 'memory' of the savannah? Would you instinctively feel at home on the east African plains, even if you've never been there?

One fascinating study suggests that we do indeed have an unconscious preference for the ancient human homeland. Researchers showed people across a wide range of ages, from eight to eighty, some photographs of various kinds of environments. They included savannah, deciduous and coniferous forest, rainforest and desert. The subjects were asked which they would prefer to go and visit. Incredibly, despite probably spending most of their time in the playground or the garden, all the eight-year-olds without exception preferred the savannah. As the subjects got older they became increasingly likely to express a preference for the forests. None of the subjects preferred the desert or the rainforest.

How could eight-year-olds be closer in their preferences to our ancestors of millions of years ago? The children's psychological make-up is bound to be closer to our innate, unsocialized mind. As we get older, the socializing effect of modern life and our life experiences exert a much greater pressure; we may, for example, become more comfortable with environments that we know or like to spend time in.

As Richard Leakey, a member of the famous fossil-hunting Leakey family, said, 'the vast majority of people who come [to Africa] feel something they feel nowhere else . . . It is the capacity homing pigeons have, salmon have, to recognize, to go back. You feel it's home. It feels right to be here.' So perhaps this preference does prove the existence of an innate memory of our evolutionary homeland, even a latent desire to return.

Surviving on the savannah

Despite its plentiful natural resources – and, in fact, because of those resources – the savannah, as we saw earlier, was a dangerous place to live. Just as we had a preference for these ancient grasslands, so did many fast and vicious predators. That all-powerful fight-or-flight response could not in itself guarantee any of our ancestors' survival. They needed far more than just an initial rapid mental and physiological reaction to the sight of a predator or other potential danger; they needed to survive the attack. Once their bodies had become primed to fight or flee, they then had to get on with the rather crucial business of day-to-day living itself – but how?

One day, in the era that we now call the Middle Pliocene some three and a half million years ago, two ape-men walked across a volcanic plain at Laetoli on the eastern edge of the Serengeti Plain in Tanzania. We can only guess where they were going or where they had come from, but we do know that journeying across these miles of open grasslands was probably part of their daily life – in fact, perhaps the only way to find food and water. The pair trailed along in a line, one after the other – just another day on the east African savannah. But it so happened that the nearby volcano of Sadiman had just spewed out huge amounts of fine ash which had covered the plain like a soft layer of grey snow. This ash contained a substance called carbonatite. When that gets wet it becomes gloopy, rather like wet cement. Earlier in the day rain must have fallen, and as the two hominids walked over the plain they left a trail of footprints in the ash behind them. Some time later a now extinct predecessor of the horse, a Hipparion, crossed their paths, leaving its own distinctive trail of hoof prints. The sun must have come out and dried the carbonatite, preserving fifty-four footprints in all, and later more layers of ash fell, burying the tracks for some three and a half million years.

In 1978, Mary Leakey, mother of Richard Leakey, was working in the area. Her team had found many animal prints but none of them expected to find the mark of early hominids. These extraordinary coinciding events – the soft volcanic ash, the rain followed by sun, and then successive new falls of ash – had preserved their footprints intact. Now they are a ghostly imprint of our ancestors, a remarkable record of a journey made an unimaginably long time ago.

What made a few old footprints so extraordinary? It was something very simple and immediately obvious to Mary and her team: the hominids had been walking on two legs. The find was the very first confirmation that as long as 3.6 million years ago there were 'ape-men' who no longer moved around on all fours, but instead stood and walked upright. They probably belonged to a species called *Australopithecus afarensis*. A few years earlier, a remarkably complete female skeleton of this same species had been found at Hadar in Ethiopia. She was named Lucy, and has since become the most famous fossil skeleton in the history of archaeology. The footprints at Laetoli were a clear confirmation of Lucy's ability to walk on two legs, a fact the archaeologists had suspected from examining the structure of her skeleton but had no way of proving from the bones alone.

So our ancestors were walking on two legs far earlier than anybody had suspected. What had triggered early humans to start moving around on only two legs rather than four? Some people have supposed it might have been similar to an advanced chimpanzee-like stance for reaching the ripest fruits from the tops of small trees. Others believe that being bipedal conferred huge everyday advantages on our chances of survival: we could cover much greater distances; being upright, we stayed cooler because we absorbed less heat from the sun (at midday, near or at the Equator, only the top of the head would be exposed to direct sunlight); we could hunt and gather across much larger and possibly more lucrative territories.

Whatever the precise sequence of events that led our ancestors to stand up on two legs, we know that an upright stance became a fundamental key to their ongoing survival and success as a species. Look at any newborn baby at just a few hours old and you'll see the vestiges of our ancient ancestors' instinct to walk. Very young infants possess a walking reflex for just a few hours, as if unconsciously trying out some ancient survival strategy. Of course they are unable to support themselves, but if held upright, you can quite clearly see the legs moving as if they were attempting to walk. The reflex fades rapidly and the child then has to progress through crawling until it can walk properly for the first time, but the instinct to be on two feet is strikingly obvious.

For the hominids, something hugely important also came hand in hand with being on two feet – in fact, it may have been one of the reasons they gave up walking on all fours in the first place: being upright meant that we now had our hands free. We were able to carry objects on these journeys across the scorched grasslands. Probably we carried food. Perhaps mothers carried babies. We could gather fruit and vegetation more effectively. But the most important change was yet to come. It took a while – around a million years, in fact – but as these early hominids evolved from *Australopithecus* into the genus *Homo*, the lineage which eventually led to modern man, they began to use their hands to make stone tools.

Tools for survival

Making tools was a key stage in the development of early man. This ability was the result of having a big brain, but tool-making in all probability was also a stimulus for its further growth. And tool-making, rather like the development of language later, is almost certainly part of man's instinct. Across eastern and southern Africa, *Homo habilis* (the first member of the genus *Homo* and almost certainly a descendant of Lucy's species, *Australopithecus afarensis*) left us some fascinating clues to his way of life in the form of scattered pieces of rock which had been intentionally broken into crude cutting or chopping tools. *Homo habilis* might have been nicknamed 'the handyman', but these tools weren't so complicated to make – just take a piece of flint and hit it hard with another rock, being careful not to crush your fingers in the process. Well-trained chimpanzees can accomplish the same feat (although chimps have never been seen to use stone tools in the wild). These broken rocks are called flakes, and they must have been pretty useful to the hungry *Homo habilis*. As a tool for butchery they're extremely effective, with a blunt back edge to grip and one sharp edge perhaps used to skin animals and cut meat from the bone.

After *Homo habilis* came *Homo erectus* (and its sister species *Homo ergaster*), spreading out across Africa, Asia and Europe from around 1.9 million years ago and lasting until the birth of the early forms of *Homo sapiens* around four hundred thousand years ago. *Homo erectus* was a vastly improved model. As the name implies, they were approaching the fully erect posture of modern humans, but what really marked them out was a substantial increase in the size of their brains, to 1000cc or more (modern brain sizes average around 1350cc, something like three times the size of any other animal when compared by body weight). This increased cranial capacity almost certainly led to a much increased cognitive ability. When it came to tools, *Homo erectus* went one step further than his handyman ancestors.

The fashionable tool for the *Homo erectus* about town was the classic handaxe – the symmetrical teardrop- or almond-shaped blade, cut from flint or a similar stone, which can be found in great numbers at ancient sites across Europe, Africa and Asia, from Indonesia to Wiltshire. It takes a lot of craft to cut, or 'knap', one of these handaxes. If you had a go yourself, you'd probably come somewhere close to the best examples of ancient handaxes, but only after quite a few months of practice. The process requires a fine appreciation of how a piece of flint fractures under impact, the optimum angle at which to hold the striking edge of the rock, how to use a softer tool like a piece of antler as a hammer, and how to trim the circumference to a tapered and neatly curved cutting edge. *Homo erectus* were clearly craftsmen, even if their repertoire of talents was somewhat limited.

The surprising fact is that once our ancestors had finally mastered the teardrop handaxe, its design remained almost identical for something close to an incredible 1.8 million years; from China to the Thames Valley, no other tools appear to have been fashioned, although some have argued that perhaps other wooden tools have not survived. Whatever the case, for the flint handaxe to remain of the same design for such an extraordinarily long period, we can only assume that it must have had an extremely important functional use in everyday life, a use that was probably vital to the success and survival of *Homo erectus*. Sadly, even with the good fortune of having evolved to the degree of owning today's 1350cc brains, we must grudgingly admit that we are not completely sure what handaxes were actually used for. In theory, they could have been used for butchery, but because they are sharpened all the way round they are more dangerous and more uncomfortable to hold than a simple flake with its one cutting edge. They could have been used for digging roots out of the ground, but the same problem exists: again, a thin-nish, shovel-shaped flake is far better and more ergonomic. So why spend time and energy making such an elaborate tool which doesn't work as well as the older version?

Researchers are rather stumped on this front, although a few outlandish theories have been put forward. William Calvin, a neuro-scientist, believes they were put into action as throwing weapons, as a kind of sharp-edged discus that could be thrown at a herd of antelope or gazelle drinking at a water hole or lake. His theory is called the 'killer frisbee' hypothesis. Calvin decided against a search for proof among early human remains. Instead, he did a bit of lateral thinking and hired a professional discus champion to try it out, and they found that with a little practice the handaxe could be a fast and accurate projectile. The sharp edge could probably hit an animal with enough force to throw it off balance, or even to topple the beast. The herd around it would panic and try to flee, scattering in all directions; the target animal, disorientated and off balance, would get trampled in the rush. The hunters would then run in and grab their prey, perhaps killing it by clubbing it with rocks.

Other authorities on the handaxe question are not convinced and think they had a far more mundane and less athletic use, like simple butchery. Instead of discus champions, they employed professional butchers to test out their theories and the butchers did declare the handaxe a well-designed tool for separating meat from bones. But this explanation does not say why they did not settle for flakes, which are quicker to make and easier to use.

What if *Homo erectus* was just a big show-off? We all know that's a particularly human trait. There may not have been a practical reason or a killer application for these elegant upgraded tools. Perhaps it was simply a display of skill, our ancestors' version of the peacock's tail, a display of virility that came about as a competitive means of attracting females. Whatever its precise use, what the handaxe does represent is the foundation of our uniquely human ability to make the environment work for us, to mould and shape it until it is fashioned for our own survival.

Man the Hunter

The most basic laws of human survival meant that our ancestors had to develop surefire ways of finding themselves food. How did they do it? We know for sure that as *Homo sapiens* was spreading throughout Europe some half a million years ago, he learned how to hunt and kill wild animals. Recently, an open-cast coal mine in northern Germany yielded up three long wooden spears carved from spruce. They were shaped like javelins, over six feet long, with a sharp point at one end. Shorter lengths of spruce were found with grooves at one end, suggesting that a stone blade had been slotted in at the top. They were found to be four hundred thousand years old, making them the oldest non-stone tools ever discovered and strong evidence of our hunting ability. But that is still relatively late in our evolutionary calendar. Before then, even with stone tools like flakes and handaxes, the picture is rather less clear.

In the 1950s, Raymond Dart, the archaeologist and professor of anatomy, discovered piles of animal bones in a South African cave. He believed that the cave, called Makapansgat, could have been a home for Australopithecines seeking shelter from the elements, maybe three or four million years ago. The bones, some belonging to ancient canines like hyenas, were broken up and marked with gouges and scratches. They seemed to have been stockpiled, and Dart thought they could only have been gathered together for one reason – to use as weapons. A large primitive hyena jawbone would have been a pretty effective stabbing implement, easily used by hominids to kill other animals, and maybe even one another.

The idea that this pile of bones was a weapons cache was seized upon as proof of man's hunting ability. Well before the invention of the stone tool, our ape-like ancestors had used these bones – which they might have found, scavenged or kept from earlier kills – for the purposes of violent hunting. Dart thought that this was the defining characteristic of the hominids, the very reason why our line had split off from the great apes. Calling it the 'killer ape' theory, he saw our hominid ancestor as Man the Hunter and believed that this hunting ability marked us out as special.

However attracted we are by the idea of Man the Hunter, the consensus is that we would have been hard pushed to compete with the brawn and speed of other major predators on the savannah such as scimitars and leopards. Most contemporary archaeologists and palaeontologists believe that early man was not an accomplished hunter. Realistically, he lacked both the strength and the speed to compete with the real players, the big cats and the hunting canines which would certainly have stalked herds of grazing animals gathering at watering holes to drink. Our flying rocks, no matter how perfectly crafted, would have made little impression on a powerful sabre-tooth.

The meat-eaters

Meat is a contentious subject. Man the Hunter, the 'killer ape', was of course supposed to be enthusiastically carnivorous, and presumably he would have had the skills necessary to catch enough prey to provide him and his family with a constant diet of fresh meat.

But what about gathering? In the past, some people have advocated vegetarianism on the basis that our ancient ancestors were themselves vegetarian. If their natural diet over thousands of years had been one consisting almost entirely of vegetables, fruits and pulses, then we too would surely be much healthier if we stuck to those kind of foods. Some extremists of this way of thinking have even concluded that if Australopithecines ate only sweet fruits, we should do the same.

While it may be true that five or six million years ago the bulk of our diet was vegetables and fruit, there are convincing clues from fossil evidence that we later became confirmed carnivores, even if we didn't kill the animals ourselves. Flakes and handaxes might well have been used for butchery, but they don't necessarily prove we caught and killed our own prey. We could just as easily have been scavengers, stumbling across the half-eaten remains of a lion kill – say, a deer or gazelle – and rushing in with our stone tool kit to cut the remaining leftovers from the bones. It's certainly not as glamorous a lifestyle as Man the Hunter, but it seems more practical and certainly more plausible.

However the meat was found, we know almost for sure that our ancient ancestors ate it. Hominid remains have been found next to fossilized animal bones with cut marks which appear to have been made with sharp stone tools. Some larger bones were broken, smashed with rocks, perhaps to get at the tasty marrow inside. And microscopic analysis of these ancient marks shows that in many cases the cut marks are laid on top of teeth marks made by large predatory animals; in other words, the big cats and other hunting animals made the kill, and we arrived to scavenge the remains.

Bones belonging to more recent hominids also yield strong evidence of carnivorous diets. A recent study led by the University of Oxford undertook to analyse some jawbone samples dating back to nearly thirty thousand years ago. They also looked at the bone composition of other animals from the same time period, including wolves, mammoths and cave bears. The researchers measured the differences in stable isotopes – the individual signatures, if you like – of carbon and nitrogen in the bones. Rather like the rings laid down in a tree trunk, our bones carry a record of the signatures of different foods we eat. The results were quite astonishing: these people's diets had been almost entirely made up of meat, and what is more, they ate almost as much as the confirmed meat-eating predators with which they were compared. Perhaps the Man the Hunter theory wasn't so far off the mark after all. Surely they wouldn't have been able to scavenge an entire meat diet without relying on a daily base of plants and vegetables?

At any rate, we know that early humans were definitely not vegetarians. Indeed, as *Homo erectus* moved out of Africa towards northern Europe and Asia, the climate would have made a purely vegetarian diet virtually impossible. In the cooler northern Ice Age climate there would have been very little vegetation, especially in winter. Modern communities who live in Arctic conditions eat an enormous amount of meat. For example, among the Inuit people, meat accounts for up to 90 per cent of the diet.

The perils of food

Think for a moment about eating meat, whether hunted or scavenged, in the searing African heat. Our Australopithecine ancestors might have prepared themselves a tasty savannah supper of three-toed horse steak and antelope tongue, but there were risks. Meat carries dangerous pathogens and parasites. Even that hominid-style salad on the side could spell disaster. Plants are full of toxins. Brussels sprouts, for example, contain a chemical called allyliothiocyanate, a slightly toxic chemical many people, especially children, find very bitter. Black pepper contains sarole, a carcinogen (in extremely small and benign quantities, so don't stop seasoning the soup). Nutmeg contains substances so toxic that eating just three of them can be fatal.

Just as plants have evolved to produce toxins to protect themselves from being eaten, we've evolved methods of avoiding them. Bad smells and tastes often indicate the presence of harmful compounds. If we inadvertently eat something toxic, we have secondary lines of defence: gagging, vomiting or diarrhoea. If you are a woman who has been pregnant, you'll probably have experienced for yourself some of our ancestors' ancient survival skills when it comes to food. There really seems to be an evolutionary explanation for morning sickness.

A fetus's major organs are developing between six and fourteen weeks after conception. At around the same time, the mother's immune system temporarily lets down its guard, giving the fetus free reign to establish itself in her uterus. As a result, pregnant women are particularly susceptible to bacteria and viruses during those weeks. Less than absolutely fresh meat, for instance, may well contain organisms such as toxoplasma, a type of parasite. Normally, our bodies would destroy such pathogens without a problem, but in early pregnancy they can cause maternal infection and even trigger a miscarriage. Over thousands of generations, it appears we have evolved a strategy for protecting the developing fetus against dangerous poisons.

Not only are pregnant women regularly sick, they also experience strong aversions to particular types of food. In a recent study of some 79,000 pregnancies in sixteen countries around the world, 66 per cent of the women reported that they had suffered some degree of sickness during the early stages of their pregnancy, and interestingly, nearly a third of them reported a strong aversion to animal products, particularly meat, fish and eggs. Once the vulnerability of the fetus is reduced after the first trimester, the nutritional value of the food outweighs the risk and the condition usually disappears. It may be pretty unpleasant for the expectant mother, but this mechanism probably played a large part in helping our ancient ancestors to have healthy children, so it got passed on to future generations. Even though we live in a generation of refrigerators and 'best-before' dates on food, the chances are your body's Stone Age instincts for survival, honed over hundreds of thousands of years, will still be as strong as ever.

Food for survival versus surviving your food

The main challenge when it comes to eating well, whether we are talking about a band of hungry *Homo habilis* or guests at a Hampstead dinner party, involves getting the right nutrients in the right amount and avoiding harmful toxins. These are the driving forces behind our culinary instincts. Unfortunately, our way of life has changed, the availability of food has been transformed and the ancient instincts which once helped us survive can now be self-destructive.

There is a very good reason why McDonald's has an annual turnover of a whopping thirty billion dollars and why 96 per cent of American schoolchildren can identify Ronald McDonald but not necessarily the Pope or the President of the United States: humans crave fat and sugar. For our savannah ancestors, the most calorific foods on offer contained the highest levels of fat (in the case of meat) and sugar (in the case of ripened fruit). Meat and ripened fruit were not that easy to come by, so the more powerful our craving for these foods, the more effort we would put into finding them. Those who carried the genetic programming for craving fat and sugar and managed to satisfy those cravings were as a result stronger, had more stamina, more reserves in times of scarcity, and better fertility. They appeared to be a better evolutionary bet all round.

But think back to the amount of meat some of these hominids must have eaten – it made up almost all of their diet in some cases. Didn't they suffer from the modern maladies we associate with eating too much fat – heart disease, high blood pressure, obesity and the rest? If you look at modern-day hunter-gatherer societies like the Nanamiut of Alaska, the Aborigines in Australia or the !Kung of Africa, there's virtually no evidence of these kinds of diseases whatsoever. Even more interesting, these people traditionally have very low blood cholesterol levels.

It turns out that wild game roaming around the savannah have a much lower fat content than the corpulent, sedentary cattle we breed to eat (about 4 per cent compared to 25–30 per cent). That explains why venison or buffalo meat is much drier and tougher than beef or pork. The fat of wild game was also about five times higher in polyunsaturated fats; modern cattle are packed full of more dangerous saturated fats. Our domestication of cattle has also provided us with a cheap source of fat-rich dairy products like milk, cream and cheese. McDonald's hit the bull's-eye with their targeting of our instinctual human cravings: a cheeseburger, fries and a milkshake offer plentiful amounts of energy-rich fat. Alongside this, our craving for sugar is now satisfied mostly by refined sugar in the form of soft drinks, chocolate, sweets, and all kinds of processed food, rather than the healthier fructose found in ripe fruit.

If you were to bump into a member of the modern-day !Kung people or a Nanamiut, you'd notice immediately that they are lean, healthy and athletic. Look around at the inhabitants of any city in any industrialized nation in the world and the difference is stark. We have an endemic obesity problem, especially in the West, and with the globalization of fast food it's now also increasing significantly in developing countries. A staggering fifty-eight million adults in the USA are overweight – not far short of half the entire adult population – and another quarter of Americans are clinically obese. We know that some 70 per cent of cardiovascular disease is directly related to being overweight and obese. Cravings for food have been transformed from being our survival instinct into a new, self-inflicted and very deadly predator.

Yet for most people, dieting is wholly unsuccessful. Many different studies have reached the same conclusion: around 95 per cent of dieters either fail completely to lose weight, or they lose weight but then put back on the same amount. And we know it's an ongoing effort: Americans spend almost the same amount as McDonald's annual turnover on trying unsuccessfully to lose that extra weight.

The fact is that simply trying to reduce your calorific intake won't necessarily help. Why? Look again to our ancient ancestors. The human body was designed for times of scarcity. Life on the savannah was marked by a high degree of uncertainty in many respects, particularly in the availability of food. In times of famine, the body had to defend its weight fiercely – a fact reflected in people's experience of dieting today. If you start reducing the amount of food you consume, your body recognizes this as an ancient savannah famine and immediately slows down your metabolic rate. Even a minimal weight loss of a pound (450g) a week will trigger this response. In other words, the less you eat, the less you need to eat before you start putting on weight again.

Imagine yourself on a 'physical activity' scale of one to ten where one is, say, Homer Simpson, and ten is a professional athlete who trains for several hours every day. For those people who have an activity rating of four or five upwards, there is a direct relationship between the amount they eat and the level of their activity. The more active they are, the more they eat; but as they go up the scale the average body weight stays pretty much the same. Their bodies are telling them to eat to compensate for the increased level of exercise, and because they are burning off this added energy they do not get fat. However, once we more sedentary specimens drop below four on the activity scale, this relationship breaks down. As early humans we probably tended to be in the four-and-above range: there wouldn't have been much free time to sit around on the savannah, discussing the movement of antelope herds over a tuber or two. There were chores to do: collecting water, gathering roots, crafting handaxes and flakes, looking after the babies. Scavenging would involve hours of walking, and running if they were hunting live prey. The optimum situation was to burn all the calories ingested. That is why, even today, raising your metabolic rate through exercise is a far more effective way of losing weight than cutting down on your food intake.

Becoming modern humans

While our craving for calorie-rich foods can be a problem today, three million years ago it had an unexpected and startling consequence. Meat provided the rich nutrients and calories that allowed the brain to increase massively in size, bringing it up to its present dimensions.

Our brain consumes one fifth of the energy that our body uses, even though it only makes up 2 per cent of our total body weight. In terms of energy, it's a very expensive organ to run, costing the body pretty much the same amount of energy as a fifteen-watt light bulb. In fact, it's rather as though humans are running massively powerful supercomputers while the rest of the animal kingdom makes do with pocket calculators.

Eating a lot of greens like a cow or a rabbit requires a huge long gut and is not that efficient. It's thought that to accommodate the new, larger, hungrier brain our digestive tracts had to become shorter and less energy-sapping. That was only possible because our ancestors were driven to eat better-quality protein and fat derived from meat. Once fire had been harnessed, about one and a half million years ago, we learned how to cook and that too made both meat and some vegetables easier for our bodies to digest. So it seems that our craving for meat led to a shorter intestine and a larger brain, triggering a revolution: the rise of the brainy hominids, a phenomenon we'll explore next.

chapter**two**

The Growing Brain

The fragile child

A few years ago, in 1996, the mild-mannered Dr André Keyser, a retired palaeontologist, was digging around with some French colleagues in the sediment in the caves at Drimolen, some forty miles from Johannesburg in South Africa. Looking for precious fossils requires patience and extreme gentleness. 'You can't imagine the emotion when you're digging with a trowel,' says Keyser. 'It's absolute fear because it's so easy to damage fragile bones.' Suddenly, they made a fascinating discovery. Having excavated fifteen feet below the ground level of a cave floor, they uncovered human remains, entombed deep in two million years of mud and sand. Two skulls were delicately unearthed and the cranial bones were especially thin and fragile. The survival of these bones is remarkable when you consider the ravages of the environment over millennia; such delicate hominid specimens probably rarely survive. But what was particularly intriguing about these skulls was that they had originally belonged to infant hominids.

No-one had previously discovered early humans this young before. The first child had been about three years old when he died and he was an early member of the *Homo* lineage, our direct ancestors. The second was a ten-month-old infant and a member of the hominid line *Australopithecus robustus*, a small-brained, heavy-set vegetarian, a line that became extinct at least one million years ago. What to me is particularly remarkable about these skulls is that the three-year-old had a cranium roughly the size of a grapefruit – about the same size, says Dr Keyser, as the skull of a modern-day human fetus.

Just because they were found close to each other in the cave does not necessarily mean they were living at the same moment in time. The fact is they could easily have lived ten thousand years apart, such is the relative crudeness of the dating process, and just happened to be entombed in the same deserted cave, sinking slowly into the silt over the next two million years.

What caused these two young children to die? Did they starve or die of cold, or were they the victims of some terrible illness or contagious disease? Maybe their bodies were pulled into the cave and picked at by scavengers. Dr Keyser thinks it likely that they had been dragged there by predatory cats.

We know that the early human child had been teething. In its lower jaw there are two baby teeth and a permanent molar pushing its way through, now frozen in time for some two million years. The one thing we can be sure of about these children is that they were undoubtedly entirely dependent on their mothers for food, warmth, shelter and protection from the vicious predators we know stalked the savannah. Infant hominids, just like modern infant humans, were completely helpless and vulnerable for many months after birth.

Let's look for a moment at a chimpanzee baby. It will have had a slightly shorter period in the womb than a human baby, around 230 to 240 days. Although the baby chimp is relatively helpless immediately after birth, within just twenty-four hours its instincts mean that it's able to crawl around quite efficiently, and after just two days it will clamber up its mother to suckle and is strong enough to hang onto her body for transportation without any assistance at all. It would take a human baby, which has only a slightly longer gestation period, maybe nine months or so to reach a comparative level of strength and control. In fact, a newborn baby left to its own devices probably wouldn't survive for much more than forty-eight hours. A human infant's vision is extremely poor; he perceives a fuzzy world without much detail. Babies have motor skills which are completely undeveloped; they can just about grip an object, they can root about for a nipple against their mother's chest, and, as any new parent rapidly discovers, they are pretty adept at crying, but that is about the limit of their obvious abilities. Almost all of the developments a human baby will experience in its first year – advances in cognition, motor skills and vision – have already taken place in a baby chimpanzee while in the uterus. We humans are born prematurely. If we were to come into the world at the same level of development as a chimp, pregnancy would take an incredible eighteen months. Compared with our humble cousins at birth, we are practically embryonic!

Given that so much of a human's development takes place after the child is born, it follows that it will be completely dependent on its parents for several years. The time and energy it takes for adults to protect and raise their babies is huge. There must be a really strong reason why humans evolved to be so helpless and vulnerable as young children. What could possibly outweigh the risks of relying on other human beings for your every need for so long? Look no further than the bigger, better human brain.

Bigger brains, bigger heads

Between infancy and adulthood, the human brain is going to increase in size around fourfold. We know that evolving bigger brains pushed our ancient ancestors forward in the struggle for survival, but there was one very serious constraint: the size of the female pelvis. There is a point at which the head will not physically fit through the birth canal. Even though the pelvis can stretch a bit during labour and the baby's head can be safely compressed a little, if the head was very much bigger than the birth canal the consequences would be disastrous. Either the baby couldn't be born at all, or serious brain damage inside the soft fetal skull would be inevitable. One theory suggests that the size of the pelvis itself was limited by the fact that we are bipedal – that we walk on two feet rather than four. The structure of the pelvis is crucial to our upright posture, and if it were much bigger it couldn't support the spine properly. In other words, if babies were born with brains and therefore heads that were any bigger, the human female would have such a wide pelvis that walking would be very limited indeed and progression on all fours would be more efficient.

As it is, and as any mother knows, the pelvis appears to be at its limits. Despite its widening over the past three or four million years, women still have to live with the pain of childbirth. No other species appears to undergo anywhere near the level of pain a human female experiences. In his book *The Dragons of Eden*, Carl Sagan points out that the connection between the evolution of intelligence and the pain of childbirth is alluded to in the Book of Genesis. In punishment for eating the fruit of the tree of the knowledge of good and evil, God says to Eve, 'In pain thou shalt bring forth children.' With knowledge comes a bigger brain, and with a bigger brain comes pain, at least for mothers-to-be.

Thus a paradoxical tension exists between the dangers of being born prematurely with what amounts to a brain that is only a quarter of its adult size and the need for our skeleton to be able to support us to walk upright. We certainly can't measure accurately the strength of different kinds of evolutionary pressures, and how they match up over, say, a million years of hominid history. We can only make informed guesses. Was bipedality really so important? Surely the large amounts of energy involved in having to care for our young must have been a powerful reason for our brains not to get bigger? There are many different ways of approaching the puzzle. Perhaps it is connected to the fact that women, if pregnant for too long, would themselves be vulnerable to predators or starvation if their family or group did not look after them. Perhaps the need for an increased brain size – with its increased cognition, memory, social skills and so on – was simply so great that it outweighed every other consideration.

Baby instincts

A newborn child can show us how human instinct – a genetic programme which will have remained largely unchanged for tens if not hundreds of thousands of years – drives us to explore our world, how we try to make sense of the objects and people within it, how we form relationships with our parents and, most importantly, how we ensure our own survival. We should never lose sight of the fact that our genetic inheritance is all about survival, and that's just as important for newborn babies as it is for the adult hunter.

As we've seen, despite being born with extremely large brain cases relative to their bodies, our babies are actually very undeveloped and helpless. Powerful instincts must step in and drive the newborn to persuade its parents to feed it, clothe it and keep an eye out for marauding beasts and wicked uncles. As it turns out, our physically helpless little bundles of joy actually have a highly sophisticated system of alarms, signs, smells and other tricks that ensure they are nurtured and cared for.

From the very first moment of a baby's life, instinct can be heard echoing through hospital corridors; the baby's cry is a crucial signal, a kind of SOS, which tells the mother to attend to the child. Crying is the first line of defence in a newborn's armoury. It rarely fails to get attention and will be the first response if the baby feels cold, discomfort, pain or hunger. If a nappy needs changing or it is time for a feed, crying will alert the adults looking after the child. Just on the off chance that they're wandering some distance from camp, perhaps to gather food, babies wail at a volume which is completely out of proportion to their body size. In fact, if their crying was scaled up to adult proportions, the decibel level would be roughly that of a pneumatic drill! And in return, there is something in a mother's instinct that makes a baby's cry almost irresistible.

Audible alarms to alert mothers or fathers are not confined to human infants. Many species employ similar strategies but birds are particularly noisy. Chicks in a nest are also relatively helpless when young, and need to be fed by their parents. They will 'cheep' until they fledge and are ready to take to the skies and fend for themselves.

A baby, when it comes down to it, is a little bundle of unconscious self-interest, and the techniques of emotional manipulation it uses to make its parents go all gooey with love, in spite of all the energy and time they have to devote to it, have been honed and refined over hundreds of thousands of years of evolution.

Pheromone fathers

It's fair to say that in almost all cultures we know of, both past and present, the mother carries out the lion's share of the work in bringing up a baby. Breastfeeding is a most important part of the job, a time during which the bonding process has ample opportunity to develop. The baby's priority, though, is to make sure the mother's breast is present and available for feeding at whatever unsociable hour it deems fit.

But fathers also play a role in keeping the baby warm, fed and happy. As we shall see in the next chapter, there are techniques employed by women to keep their man from straying; after all, there is little doubt that historically men have been important providers of food and security for the family group, especially if the mother has a baby in tow and is therefore more vulnerable to attack or potential starvation. But it turns out that it's not only mothers who have tried and tested methods of keeping their men on a tight rein. New research from Sweden shows that newborn babies have evolved an even more wily technique to attract their fathers: smell.

In the study, a group of newborn babies, just a few weeks old, and infants aged between two and four were gathered together. The experimenters gave each child a thorough bath with unscented soap, then dressed them in plain white T-shirts. The T-shirts were then left on the children for a couple of days so that sweat could accumulate. After that, the little infants were relieved of their duties. The T-shirts were taken away to a lab to be sniffed by a group of adults, both men and women, some of whom were parents, some of whom were childless. Each adult smelled the unlabelled T-shirts in groups of three: the baby T-shirt, the infant T-shirt, and an unworn T-shirt as a control sample. Each was then asked which T-shirt smell they preferred. The results were rather surprising. The women could not tell the difference between the baby T-shirts and the infant T-shirts. In fact, most of the women said they preferred the smell of the unworn T-shirt. The men, however, particularly the men who had children of their own, stated a distinct preference for the newborn baby T-shirts.

Odours can be incredibly powerful mental stimuli. We've all experienced a sudden and sometimes emotional rush of memory after smelling something we recognize from our childhood – candyfloss, asphalt or the smell of a favourite pet. Our brains may have lost a large proportion of our smelling finesse, and compared with many mammals we are amateurs in the smelling stakes, but we can still pick up and recognize, whether consciously or unconsciously, tiny amounts of odour-making chemicals in the air.

Babies, it seems, have a strategy which takes full advantage of this. After analysing the chemical make-up of the newborn T-shirt smell, the researchers concluded that the smell included pheromones, chemicals which aren't smelled consciously in the normal way but are transmitted directly to the emotional centres of the brain through a special bundle of nerves in the nasal septum rather than via the olfactory nerve, the way conscious smells are usually transmitted to the brain.

Could this 'smell trap' have evolved to make sure the fathers stick around and help provide for the child? Possibly, but there may be another explanation. Babies are sensitive to violence and physical danger. In fact, we can see the very first signs of the emergence of the fight-or-flight response in very young babies. It's called the 'Moro' reflex. If the baby hears a loud noise, it will throw its arms forward, bring them together in a protective gesture and then relax. A few weeks later this instinct develops into the 'startle' response: the mouth opens (possibly to increase oxygen flow), the head lowers and the shoulders and arms sag, as though the baby is expecting a body blow.

Could it be that the baby pheromones are being used to stem violence and aggression among the adult males around the infant? Karin Bengstson, who headed the study, believes they do have a calming effect, counteracting the aggression that derives from the traditional male role of hunter and fighter. From the baby's view of the world, male aggression and violence are dangerous; babies much prefer cuddles and comforting to shouting and throwing. Perhaps one day these special baby pheromones may even be used to create an artificial baby aroma to calm violent and criminally aggressive men.

Babies and faces

Human infants aren't content with just making noises and smells to alert the world to their needs and desires. Very soon after birth, babies start to imitate the facial expressions of adults. If an adult smiles, babies will start to smile back. If the adult frowns, the baby will also display the beginnings of a frown. The parents will find this endlessly fascinating (although no-one else will). Incidentally, I recently tried this with some newborn babies born at Hammersmith. One of the television producers of a BBC programme called *Child of Our Time* wanted me to pull faces at infants just a few hours old, so they could film the babies imitating me. I was a total failure – perhaps it was my moustache. But I did discover a new-found talent. Every one of the babies I picked up out of its cot and grimaced at, immediately fell asleep. The mothers were all most impressed, but the TV crew got very impatient. I only wish I had discovered this talent many years ago when Tanya, my eldest child, was a very fractious baby.

Even though it didn't work too well with the very young babies, just a few hours old, with whom I experimented, there is no doubt that with this instinctive trick, babies can exert an extraordinary emotional pull on their parents. It is central to the bonding process – the growth of the deeply rooted, powerful emotional ties between parent and child. At the heart of this process is a baby's capacity to recognize the human face.

Newborns appear much more interested in a face than in other random objects, and they lock on with increased interest to the basic arrangement of two eyes, a nose and a mouth. Bearing in mind their poor vision and extremely limited understanding of the physical and human world, this ability is pretty remarkable. But newborn babies almost certainly can't grasp the significance or meaning of a face or an expression – that doesn't begin to develop until they are about six months of age. It is purely instinct which programmes them to respond in the way that they do, and it is this reaction which means the parents are suckered into caring for and protecting the baby with far more love and attention than they would otherwise offer. We project our own love and emotional response onto the smiling face of an infant, who is simply following a genetic instruction.

This may not seem to be a romantic view of the first magical moments between parent and child, but very soon the process of attachment does allow the infant to begin to recognize its parents and discern between them and strangers. Babies do begin to respond with more interest to the adults who care for them, feed them and protect them. They're nothing if not practical.

It has long been believed that the ability to recognize different human faces is a deep-seated, innate part of being human. It seems that there is a specialized region of the brain which has evolved to control this one function. We know this through studies of particular types of brain damage in adults. Sometimes brain damage results in the patient having difficulty with face recognition, while leaving intact the ability to recognize inanimate objects or animals. Patients with damage to a different area of the brain may recognize people but not objects, suggesting that there are specific brain sections to deal with each.

In fact, we know where that special area is when it comes to recognizing faces. You'd find it just behind your right ear. And if you were unlucky enough to suffer from a rare condition called prosopagnosia, that part of your brain would be affected. The condition destroys your ability to recognize family and friends, even to remember new faces. People who actually experience prosopagnosia and those around them are startled by how central this ability is to everyday life.

Cecilia Burman is a young woman with prosopagnosia. She describes eloquently what it is like to be face-blind, how a person with this difficult condition may easily be thought to be half-witted or mentally deficient. The difficulty, of course, is being able to engage in the kind of social activity most of us take totally for granted. Because to someone with prosopagnosia a person's face is meaningless, aspects of his or her clothes, voice, hair, even the hands, can be crucial to recognition. She says that many face-blind people actually memorize the clothes of the people around them so that they can recognize them during a day's work, or at college. Greeting people is a real problem because, of course, if you do not immediately, within a few seconds, interact with an old acquaintance on meeting him or her anew, you are quite likely to have lost a friend for ever. Consequently, face-blind people often pretend to know people they've never met before and go to great lengths to avoid using names, or ask leading questions about what they've recently been up to. In her really interesting website (www.prosopagnosia.com), Cecilia Burman compares her strange predicament to that of normal people looking at a few inert stones. She shows photos of different stones, and after giving them names – Fred, Daniel, Matilda – she describes how each can be recognized by individual features: a bit sticking out on the right side, a smooth polished surface, a crack down the centre, and so on.

Face recognition is at the very heart of our social lives, in the most fundamental sense. Being human is about living with and forming relationships with others. If you cannot recognize those around you and those close to you, that unique and very special window to that world is missing.* Most of us can, thankfully, recognize dozens, if not hundreds, of faces. We are also experts in interpreting expressions, particularly infants' facial expressions. This is another adaptive mechanism which means the infant will essentially be less vulnerable in the first year or two of its life. Adults can tell if babies are happy or shy, scared or hungry.

Remarkably, research has shown that the reverse is also true. Relatively young babies can read our emotional expressions almost as well as we can read theirs, and if the mother appears to be scared

* Interestingly, tests show that when people suffering from prosopagnosia are brought together with strangers there is a significant change in their skin conductance, while no change in conductance is measured when they see people they should 'know'. The response is controlled by an unconscious part of the brain, the limbic system, and it suggests that unconsciously the patients 'know' which faces belong to strangers and which belong to friends, but are simply unable to access that information.

– owing to an unwelcome intruder, a dangerous predator or even an exposed cliffside – then babies take notice. They can even monitor the tone of an adult's voice for an indication that something is wrong. In one rather mean experiment, mothers smiled at their babies while simultaneously talking to them in a frightened tone of voice, and vice versa. These conflicting signals clearly did get recognized by the infants because the poor mites were thrown into confusion and got extremely agitated.

Playtime

Human infants have evolved another almost universal instinct which helps them out in the world. You may not think that children's make-believe games and wendy-houses have much to do with their everyday survival, but research has shown that play is a very significant and deep-rooted instinct with an important purpose.

Take a look at the animal world and you'll see that playfulness is actually quite a rare trait. It's only commonly found among mammals, although a few of the more intelligent birds like magpies and crows are occasionally observed playing. Play can even be quite dangerous for some animals: 80 per cent of deaths among young fur seals occur because the playful pups fail to spot approaching predators. And while human infants don't usually come to any harm with a Barbie doll or a Lego set, it's been estimated that young children spend up to 15 per cent of their energy cavorting about. In evolutionary terms, that's fiendishly expensive. There must be a good reason for this activity to take place, other than just for the sake of good fun.

A new study carried out in Australia has revealed an interesting link between brain size and playfulness. The researchers compared fifteen different orders of mammals, from dolphins to rodents to marsupials, and showed that the animals with bigger brains (for a given body size) spent significantly more time playing. And when the team narrowed their gaze to a specific group, the marsupials, they discovered that the wombat was far more playful than its lazier and much smaller-brained cousin, the koala.

When we watch children play, we see that they're often practising adult behaviour. A one-year-old will show empathy in his or her play, and by the age of three, children also actively engage in what is called 'pretend play', imitating adult roles and projecting beliefs, emotions and actions onto individual dolls or figures, sometimes even imaginary ones. This hard-wired instinct to 'try out' life in the adult world is an important means of communication and practice. It seems that because our bigger brains need time and experiences to develop in those first few years of life, one of evolution's solutions is to mould our emotions and understanding of life through play, so that we are as well prepared as possible for survival in the complex adult world.

The roots of learning

Toddlers are like sponges when it comes to learning. The human mind is primed to acquire rapidly the skills we need to move around in and manipulate the physical world, as well as to form relationships with other human beings and begin to communicate with them. All the cognitive talents that appear in infancy and childhood – face recognition, spatial awareness, the ability to speak and understand language – depend to a greater or lesser extent on the environment and the people with which the infant interacts.

Our astonishing ability to learn these skills has produced one of the most fiercely defended theories of evolutionary psychology. It suggests that we're born with minds that already contain complex psychological mechanisms, or 'modules'. In other words, the mind is rather like a Swiss Army knife: it has ready-made tools each of which has a clearly defined purpose, but rather than to dig stones out of horses' hooves, our minds are built to understand the fundamentals of grammar, or to be scared of snakes and spiders. One supporter of this theory is Noam Chomsky, the famous language theorist. He believes that we're all born with innate structures in the brain that allow us to learn language, or at least all of us have the genetic programming that will create these structures as we grow from an embryo into an infant. He suggested that language in every culture has a 'Universal Grammar' – a similarity in the way in which sentences in all languages are constructed – which reflects this inborn psychological 'module'.

Babies do indeed seem to like the sound of voices. Newborns display more of an interest in the sounds of spoken language than, say, the sound of a bell ringing. But how can we tell how much attention a newborn child pays to anything at all? Surely they are too busy crying, feeding and defecating? One answer is to use an artificial nipple that contains a small sensor to gauge how hard the infant is sucking. When the sound or sight of something arouses the baby's attention, the rate of sucking goes down or stops. It is one ingenious method that can tell us something about the development of the baby brain. Infants also pay more attention to their mother's voice; it is believed that they 'tuned in' to its timbre and tone while in the womb.

A new Finnish research study has shown that, astonishingly, babies are primed to learn to recognize the sounds of new words even while they're asleep. Some forty-five newborn babies were split into three groups: each of the babies in the first group was played a tape of his or her mother's voice chanting a set of unusual vowel sounds through the night as they slept, another heard a tape of commonly heard vowel sounds, and the third was left to sleep in silence. When tested the morning after, the babies in the first group were the only ones to display brainwave activity which showed that on hearing the unusual vowel sounds they now recognized them as familiar.

Consider also some further fuel for the idea of an inbuilt language 'module' in the brain. If particular sections of the brain are damaged, through injury or a stroke, the effects on language processing can be very precise and bizarre. Sometimes a person can lose the ability to find the right words. They will constantly be grasping for even the most basic nouns and verbs, or even names, but they will have a perfect grasp of language structure: 'My, uh, you know, I need . . . uh, what's the word, ticket, for the thing, you know, to, uh, I forget the name.' Other aspects of their memory are generally not affected. Another person with damage to a nearby region of the brain will have exactly the opposite problem. They'll have instant recall of vocabulary, but they won't be able to put the words together properly: 'Want ticket train Edinburgh.' They will be hopeless at syntax and grammar, and unable to construct anything but the most basic sentences.

If we are to believe the modular theory, we are all hard-wired for a wide range of human instincts and forms of behaviour shared by the whole of *Homo sapiens*, from Aboriginal trackers to Wisconsin prom queens. The basic shape of these tools, which are constructed during the development period as a fetus, and later as a newborn, are the same for everyone. They don't depend on cultural differences or the environment in which we bring up our children. There may be differences in the timing – certain cognitive abilities may show themselves earlier or later, depending on how much stimulation a baby receives from the people around it – the modules may not even be switched on at all, but they are there, a direct result of our evolutionary history, the psychological solutions natural selection has found to the problems of life on the savannah. They have been described by John Tooby and Leda Cosmides from the University of California, Santa Barbara, as keys which have been made to fit a particular lock. The solution must coincide with the adaptive problem it was designed to solve; to take an obvious example, our fear of heights is perfectly matched with the dangers of falling off a cliff. If you think about it, modularity theory suggests that if we were hypothetically able to transport one of the more recent hominid's newborn infants into our own modern advanced world, the child, with its savannah brain modules, would probably be just as well primed to take on language, social graces and cognition as one of our own.

However, modularity theory has provoked a heated debate. The evolutionary scientist Paul Ehrlich, based at Stanford University in the USA, is convinced that the evolutionary psychologists are wrong, on the basis that the brain is simply too complex to be encoded in such a specific and detailed way.

The brain has twenty billion nerve cells and an unimaginable number of connections between them that make up our neural networks. It's been estimated that there are a thousand million connections in a piece of brain the size of a grain of sand. Ehrlich points out that even if every single one of our thirty thousand genes were dedicated to wiring up our brains, each gene would need to hold the blueprint for hooking up somewhere in the region of one billion neural connections. This cannot possibly be the case; each gene cannot hold anywhere near that amount of information. In addition, our genes must also carry a great deal of information contributing to the assembly and running of all our other complex organs. They cannot all be devoted to the development of the brain. Ehrlich believes that this 'gene shortage' is why babies are born in such a helpless state. He is convinced that we have a great deal of plasticity in the brain and that our interaction with people, culture and the physical world is an essential part of 'constructing' the brain. Our genes provide guidance and some basic instructions, but the world around us picks up where our genes cannot, causing the connections in the brain to form as we progress from embryo to infancy and beyond.

Ehrlich's view of the human brain is that of an old, rambling house which has been built, modified, redecorated, partly knocked down and rebuilt by a succession of inhabitants over a period of time. Its useless corners, uneven gables and shoddy plasterwork are signs of its quirkiness, its flexibility and its individuality. The modular theory of the human mind is more like a hi-tech prefab that arrives on site ready-built, according to a detailed plan and specification. There are no follies in the garden or temperamental window frames. These houses are energy efficient, meticulously designed, and they are all pretty much the same.

These two models of the human brain have an enormous influence on how we view instinct. If we are primarily influenced by the environment and by the learning process, then perhaps our human natures are not completely universal; maybe an Inuit child's instinct will develop differently from that of a child in Essex. If the modular model turns out to be correct, then we can be fairly certain that the mind of *Homo sapiens* is constant for both Inuit and Essex girls.

With lines drawn and trenches dug, the battle will be a long-drawn-out and bloody affair. The modular theory is at the heart of evolutionary psychology; if we are to believe it, it would seem that natural selection has been busier than we could possibly have imagined, honing and refining our mental toolkit so all that remains is for our psychological modules to be switched on at the right time. That suggests that our mind is much more of a finished product than people such as Ehrlich are prepared to admit. Ehrlich and other like-minded colleagues are hoping to leave more space for the world around us to mould and fashion our mental make-up.

The growing brain

I am tempted, at this point, to believe Emerson Pugh, the physicist/philosopher, who suggested that if the human mind was simple enough to understand, we would be too simple to understand it. However, it's certainly universally accepted that compared with our hominid forefathers, modern *Homo sapiens* has reached unprecedented peaks of intelligence in terms of our cognitive skills and abilities – and, indeed, in the size of our brains. Which brings us to one of the most pressing and hotly debated questions of all when it comes to brains: what was the crucial trigger that caused our ancient ancestors to evolve ever bigger brains?

The minds of those distant hominids are an enduring mystery. We might occasionally have the good fortune to excavate old skulls from riverbeds and from deep layers of cave sediment, but what can those finds really tell us about the day-to-day lives of our ancestors? Just as the modern medical world struggled to understand the workings of our own brains until advanced imaging techniques like MRI and PET had been developed, those troubled researchers trying to glean information from hominid skulls had even less to go on. There was no flesh, no brain tissue to look at, simply a tough old bone case. Working out the structure of the mind of a three-million-year-old hominid from a fossilized cranium is a bit like trying to recreate a symphony from the plastic case of a compact disc.

In some cases, researchers could glean some very basic information about the structures within the brain. The marks and grooves on the inside of the cranium show the paths of blood vessels as well as faint indications of the physical make-up of the different parts of a primate brain. Skulls of a few early humans, for instance, bear the traces of a structure in the brain called Broca's Area, which in the modern human brain is the centre for speech.

Generally, though, the thick protective gauze that surrounds the neural tissue will have prevented almost all details from being imprinted onto the inside of the skull. The fact is that we still don't know how the hominid brain was wired up, nor can we be clear about the complexity of its neural circuitry. And while we might have stumbled across evidence of Broca's Area in those few rare skulls, how can we be sure that in early humans it was used for the same purpose as it is now? You may think our whole quest to understand the hominid mind is somewhat scuppered, but before we get too depressed and give up the whole enterprise, there is a chink of light in the darkness. Fossil skulls are not so useless after all, because even knowing the size of hominid brains turns out to be a piece of information that is extremely illuminating.

We know that the brain would have fitted snugly inside the cranium. Casts can easily be made of the interiors of fossil skulls and it is a simple job to work out the cubic capacity of the brain from them. Those casts tell an extraordinary story: the tripling in size of the human brain, an enormous and unprecedented evolutionary leap. Poor old *Homo habilis* might have had to make do with a 500cc brain, but by the time the very first examples of early modern *Homo sapiens* came along less than two million years later, the brain was now a vastly improved model measuring at its largest a whopping 1500cc or so. What on earth could have driven this vast expansion? That extraordinary growth of the brain defined our evolutionary path; it is at the very heart of a five-million-year-long story, our time spent evolving from our chimp-like ancestors through the hominid line to modern *Homo sapiens*.

Brain and body size

For most of the animal kingdom, brain size is related to body size, with good reason: the larger the organism, the more processing power is needed to control and regulate even the most basic physiological processes, such as the circulatory system or digestive tract. It also needs to provide the cognitive capacity for motor skills and the programming for instinctive behaviour to guide mating or feeding. A sperm whale, for instance, has a brain that weighs somewhere in the region of 8,000g, whereas a human brain weighs about 1,500g, but that does not mean the whale displays any signs of great intelligence. It needs that enormous mass of neural tissue to regulate the largest animal body in the world. After all, its body does weigh around thirty-seven tonnes.

Human beings are clearly not typical. Our brains account for 2 per cent of our body weight, three times larger than one would expect for a primate of our size. Even Charles Darwin clocked that you had to take body size into account, writing in *The Descent of Man* in 1871: 'No one, I presume, doubts that the large proportion which the size of a man's brain bears to his body, compared to the same proportion in the gorilla or orang, is closely connected with his mental powers.'

But it wasn't until some twenty years ago that Harry Jerison of the University of California, Los Angeles, developed a concept called the encephalization quotient, or EQ – a measure of brain size relative to body size. It turns out that a domestic cat has an EQ of 1:0 – it has just the right size of brain it needs to control its body, making it an excellent instinctive predator. Dogs come in at an EQ of around 1:8 (as the journalist Benjamin Mee points out, 'giving them a bit of spare brain to think with'). The chimp scores around 3:0, while humans weigh in with a monster encephalization quotient of 7:4.

Jerison set about tracking its changes through evolutionary history. What he showed was that the brain's evolution had involved incredibly long periods of constancy punctuated by dramatic bursts of growth. In Jerison's studies, dolphins were the only animals that came anywhere close to the human brain size:body size ratio – the bottle-nose has an EQ of 5:6. What is interesting is that all these animals with higher EQs tend to live in complex social groups with much interaction with other members of the same species.

Dolphins are believed by many people to be very intelligent animals; some researchers, who are attempting to teach them rudimentary sign language, suggest that they may even be as intelligent as chimpanzees. But to be a little uncharitable for a moment, we'd have to say that jumping through hoops and clicking to one another underwater do not stand up as evidence for their being capable of human-like cognitive feats. Their supporters may argue, however, that other dolphin-like cognitive tricks we can't possibly understand are evidence of their equal intelligence.

Size isn't everything

The easy answer to *why* the brain tripled in size is simply that clever is better. Natural selection prefers clever people, because clever people hunt better, survive for longer, reproduce more successfully and are more likely to ensure the survival of their kids. But does bigger invariably mean better?

Let's think about the modern measure of intelligence, IQ. The physical size of our modern brains does not necessarily correspond to intelligence, at least not in the form of the IQ score. I should stress, however, that IQ, or intelligence quotient, is no longer commonly considered an objective measure of intelligence. Recognized as culturally biased in favour of the Western intellectual and logical tradition, it seems that with training you can increase your score considerably. Few experts can even agree what intelligence actually is, or how to define it. However, as a rough guide to our capacity for logical and abstract thought, IQ can be useful. The idea that brain size was directly related to intelligence was widespread among neurologists and anatomists at the dawn of the twentieth century, but the theory began to be questioned when it was discovered that the brains of several highly distinguished people of the time, who had very generously bequeathed their bodies to science, showed no outstanding characteristics whatsoever. In fact, they were all disappointingly ordinary. More recent studies have shown quite conclusively that people with particularly small brains have, for the most part, relatively normal IQ scores. In fact, even people who had brains no bigger than 750cc, about the size of the brain of an average adult *Homo erectus*, were found to have no discernible drop in IQ.

Consequently, intelligence *per se* does not necessarily call for a bigger brain. In any case, being clever alone does not automatically set you up for success on the savannah. IQ tests consist primarily of logic problems, word games and mathematical puzzles, none of which figured too highly in the lives of our ancestors a million or two million years ago. But other studies have suggested that there was another important factor. Size isn't everything after all, it seems, because there was also a dynamic change in the actual organization of the brain's architecture over time.

A theory suggested by Terrence Deacon of Harvard University sets out the argument that, contrary to conventional wisdom, overall brain size is not the single most important difference between our own intelligent brains and those of our less brainy primate cousins. Rather it is the out-of-proportion expansion of an area of our brains called the prefrontal cortex and its relation to other parts of the cortex. The cortex, you'll remember from Chapter one, is the most recently evolved section of the triune brain and is responsible for the higher-order types of thinking and planning we associate with the very essence of humanness. If we were to scale up a modern ape's brain size to the correct proportions for a human body, Deacon believes we'd find that the prefrontal region in humans would be over twice that of the scaled-up ape. More importantly, this scaling-up is not just in terms of the actual size of the cortex, but in its complexity and the degree to which it is connected to and controls other regions of the cortex. In other words, the prefrontal cortex began to branch out to the places other parts of the brain couldn't reach, affording it and its higher functions with a new dominance in the evolving brain.

Whatever developments were going on inside the brain of our ancestors, the incremental increase in brain size was some sort of advantage for the early human. We know that those who had slightly larger brains were more able to survive and reproduce, but exactly *why* they were at an advantage is the million-dollar question. What actually was the result of the growing and changing brain? What was the incredible driving force of selection that it delivered, which set our ancestors down this path of rapidly evolving intelligence? It is a problem that has stumped anthropologists, neurologists and palaeontologists ever since Darwin first suggested that we had descended from stooped and rather stupid ape-men. We still need somehow to account for the disproportionate increase in brain size over the past three million years, a time during which the brain tripled in capacity. This simply can't be explained away as a random event, and we need now to consider a range of possibilities to find some likely answers.

A brainy radiator

Let's look at a more practical theory for why the human brain might have expanded so quickly. Think back to the early hominids, struggling for survival on the baking east African savannah. Brain tissue is very susceptible to overheating; an increase in temperature of even four or five degrees Celsius can cause major malfunctions in the neurons. When the American anthropologist Dean Falk remembered that an automobile's engine was limited in power by its radiator's capacity to cool it, he applied this thinking to the human brain.

The human brain, like the automobile engine, must be kept cool if it is to function well. It follows that if the brain of an animal is not functioning well, the body that brain controls will not perform well either. Overheated brains, then, are a road to extinction in the highly competitive natural world. The evolutionary development that could have led to this advantage was a more extensive network of emissary veins inside the head, permitting more dissipation of heat. This, in turn, might have allowed the evolution of larger brains and dominance by *Homo sapiens*. Some readers may think it intriguing that human male baldness also confers considerable cooling efficiency and may be setting the stage for a new expansion of the human brain (pity about the ladies).

More seriously, did the better blood-cooling system develop in response to an enlarging brain, or vice versa? Even more seriously, it is simplistic to say that an organism just went ahead and evolved this way or that way. A bigger brain requires not only more cooling but a bigger skull, more connections and additional infrastructure. Saying simply that 'a larger brain evolved' obscures the fact that many different, inheritable changes had to take place in synchrony.

If the expanded brain really did serve as a special cooling system, there could be an interesting and advantageous side effect. A larger brain would give us more 'redundancy' in our cognitive systems. The improved models of the early human brain would have stood out as being rather like the space shuttle: if one system failed, say the control of the booster rockets or the hydraulics which operate the landing gear, another part of the system could be used instead, for example, a manual alternative. Similarly, if one part of the brain got to a dangerous temperature, close to overheating, then another, unused, area could take over; or if one cluster of neurons was damaged, perhaps the ones involved in vision or motor skills, others could be drafted in for the same purpose.

This is worth considering, but I get the feeling that the cooling theory is rather implausible. Brains are hugely expensive organs, not just in terms of the energy and nutrients needed to build and run them, but also in the number of genes needed to control their development. The whole project would take a much longer period to evolve, making it more likely to be ditched in favour of a better solution. Surely there would have been simpler ways of cooling off had it been that important. Why not, like dogs, have a bigger tongue? Panting is a simple and efficient method of losing heat. And tongues certainly don't take as much energy as brains to run.

Brain food

We know that the modern human brain is one hungry beast. It takes up some 20 per cent of the available oxygen and energy in our blood – extraordinary for an organ accounting, as we have seen, for only 2 per cent of our body weight. Its greedy energy requirements work out as being ten times as much for its size as any other organ in your body. Moreover, the brain is an engine that runs continuously day in, day out, but without the added luxury of a petrol tank. The brain isn't able to store any energy, so in order to ensure its smooth operation the body must seek out a new and inexhaustible supply of energy as part of its daily but vital chores.

Some researchers have supposed that for early humans to have evolved bigger brains, they must first have had to sustain ever higher metabolic rates. In other words, they needed to find themselves a substantially more nourishing protein- and carbohydrate-rich diet. We've seen earlier that our ancestors' drive to find and eat meat and their use of tools and fire to prepare it meant they were getting richer diets and thus more energy from their food. Could that very fact have been what was responsible for triggering the remarkable growth of their brain?

We must take care here. While it's true that rising energy levels did clearly provide the sustenance for the evolution of the brain, we certainly can't immediately draw conclusions that they were therefore its cause. Just because an animal has the energy to grow a larger brain, that doesn't mean it necessarily will do so. Why not use that energy for another purpose, like growing a larger body, for instance? Better diets clearly played some sort of contributory role in the story of our evolving brain, but there must have been another, more important factor.

Expertise

So how else can we explain the large brain of *Homo sapiens*? One theory suggests 'expertise' was a key factor in exerting the evolutionary pressure on our brains to expand in capacity. Acquiring more skills meant a better chance of survival on the savannah, but more skills meant the need for a bigger brain.

Consider the brain of a modern human who has mastered a particular skill. To play the violin to a high standard requires stunning dexterity and co-ordination, of the left fingers in particular (the hand that frets the notes), as well as highly developed co-ordination between each hand, in order for the bowing to synchronize with the fretting. If a top violinist is placed into an MRI scanning machine, we can see that a much larger area of the brain – the right primary motor cortex – is devoted to his or her left fingers when compared with a non-violinist. Two or three times as large, in fact. Violinists also have more connections between the two sides of the brain which account for the better co-ordination they have between each hand compared with a non-violin player. Similarly, piano players have a greatly increased number of neural circuits in this cortical area of the brain.

But how could a hominid practising a hunting technique pass on that improved ability to his children via his brain? Learning particular skills could not have had a direct effect on their own genetic code, nor of their descendants. Even if certain areas of the brain did increase in size during a lifetime of practising a particular skill, this change would never be passed on to the next generation. The mistake made by those nineteenth-century biologists such as Jean-Baptiste Lamarck must be avoided. They thought giraffes had long necks because they stretched up to eat the leaves from the trees, thus causing their offspring to be born with slightly longer necks. But this is not how natural selection works. A longer neck would have initially been a random mutation, a genetic accident, which meant that that individual giraffe had a slight advantage over its contemporaries. If on balance that individual giraffe was then more successful at surviving and reproducing, so the genetic mutation would be more likely to spread throughout the population.

Homo erectus was no violinist, and in those days all the ivory still belonged to the elephants, but it's clear that we did acquire skills which were used for finding food, gathering, hunting, scavenging and tool-making – all skills which would have required an increasing amount of mental real estate.

If these cognitive skills acquired by our ancient ancestors during their time on the savannah are really hard-wired into our own genetic codes, we ought to be able to find them embedded deep in our minds. One way of investigating these skills may be to examine the differences between male and female cognition. Countless behavioural studies have shown that men are significantly better at spatial ability tasks like geometry, map-reading and finding their way though mazes. They seem to have an innate talent for imagining and manipulating space in their minds which tends to play itself out during arguments over map-reading in the car, or even contests over who can reverse-park with most ease. Not exactly a survival-enhancing skill any more, but could it be that this difference between the abilities of men and women (who tend to do better with other skills such as visual memory) links us to the particular skills our hominid ancestors picked up hundreds of thousands of years ago?

Evolutionary psychologists point to an ancient division of labour. Men developed better spatial ability skills because on the savannah they did almost all of the hunting. If you have a better idea of where you and your prey are in three-dimensional space, you're going to put more meat on the table, so to speak. To hunt successfully, you've got to cross long distances over sometimes perilous terrain and stalk fast-moving prey without getting stalked yourself. Women, on the other hand, who were thought to have done most of the gathering of nuts, fruits and berries, might have developed skills that helped them to remember the specific location of these foods. To recall and find the tree which produces the best fruits each year, you must have a well-honed visual memory. These cognitive differences, though, are not necessarily genetic; they could simply come down to practice and learning. If men and women have different interests and lifestyles, for instance if men drive more often, then they will develop particular kinds of spatial awareness and expertise. Those who recoil at this rather reactionary portrayal of Man the Hunter and Woman the Gatherer are not alone. As we shall see, gender differences get more contentious, especially when it comes to sexuality.

Whether or not men and women learn or are wired differently for spatial skills, we know for sure that humans certainly have an impressive array of inbuilt cognitive abilities. Newborn babies, as we've seen already, have a selection of powerful instincts – the ultimate proof that skills come hard-wired into our modern human brains and are inherited from our ancestors. But we still have a problem to solve, a kind of chicken-or-egg conundrum: which came first, the increased brainpower or the new cognitive skills? How could we have the capacity for new skills without a bigger brain, and why would we bother with a bigger brain without having any use for it?

Let's go back to the original question: what selective advantage would there have been for people with slightly bigger brains?

The Baldwin effect

In 1896, the American naturalist James Mark Baldwin wrote, 'if the young chick imitated the old duck instead of the old hen, it would perish; it can only learn those new things which its present equipment will permit – not swimming. So the chick's own possible actions and adaptations . . . have to be selected.' He proposed 'a new factor in evolution' whereby acquired characteristics could be indirectly inherited. William Calvin describes this 'Baldwin effect' as 'form follows function'; at its core, it says that a change in behaviour can lead the way, and the genes follow.

Calvin uses an everyday example to explain it. Let's for a moment imagine the genetic code as an old cake recipe. A chef has had this recipe for years, but he doesn't need to read it any more, he can make the delicious cake from memory, and after tweaking and improvising the ingredients over the years, adding a bit more chocolate here and a bit less flour there, the end result is tastier than it's ever been. Everyone who tries the cake asks for the recipe and copies down the ingredients taken from the chef's kitchen book. Sometimes, though, errors creep into the copying process. Perhaps they write ounces instead of grams, four eggs instead of two. Whatever the reason, most of these errors (the mutations that crop up in genes) are discarded as being no good – either the cake doesn't work at all or it just doesn't have the tasty finesse of the original. The aberrant recipe gets put to the back of the pile and doesn't get made again. Sometimes, though, the errors happen to match up to the tweaks and modifications made by the original chef, thus by complete chance making the cake taste even better. These mutations are a great success; anyone who samples the cake will also ask for the recipe, and these 'useful' errors will be passed on to a whole new generation of cake-eaters. Slowly, then, after many copies have been made and many errors have crept in, the recipe will converge towards the tweaked version made by the original chef, because *this chef's cake was the best possible version.*

So there is a discrepancy between the original recipe (the genetic programme for the development of the brain) and the cake made by the original chef (the behaviour of an individual animal). By chance, a particular animal is capable of some evolutionarily useful skill, so the discrepancy has a positive effect (the cake tastes better, so more people ask for the recipe). Perhaps it is an ape-man with the ability to stand upright, or to grasp a rock better, or even to use the rock to dig out roots from the ground.

As with the cake recipe, the genetic code would have lagged behind, but useful mutations which reinforced that initial behaviour would have enhanced survival and would have been passed on to the next generation. Slowly, with more mutations, the genetic code catches up and provides a genetic basis for the new behaviour. By this way of thinking, the initial valuable behaviour, be it making a tool or walking upright, is the trigger allowing a bigger brain to become useful. Bigger brains were selected for, and we became ever better at walking upright or using tools. So a continual feedback was sparked between the behaviour, the brain, the genes and, most importantly, the survival of our ancestors.

The Baldwin effect is one possible solution to the problem of the sudden appearance of a larger brain. Larger brains certainly wouldn't have appeared for no reason, especially bearing in mind the enormity of the increase in size of the hominid's. But could an entirely non-biological, culturally transmitted form of behaviour, such as picking up a rock and throwing it, really have been the spur for the vast increase in brain size?

Killer apes and killer frisbees

Having a larger brain *per se* isn't absolutely necessary for an animal's survival. Many other species have managed to survive without a massive expansion in brain-to-body ratio. The truth is that we need only so many brain cells to be a successful hunter, gatherer or scavenger, to reproduce prolifically and to adapt to changing climates and environments. Until mankind began to wipe them out, many top mammal predators such as the big cats and great apes were extremely successful. They are extremely good at hunting down and overcoming prey which in some cases have developed sophisticated methods of evading or combating the threat. It seems that all except *Homo sapiens* managed without an outsized brain.

Fair enough, you may say. Humans, even the stronger and sturdier ones like *Homo habilis*, did not have the strength or speed to match the top savannah predators. Instead, there was something else that set the human way of hunting apart – the use of tools. Because we were the only animals to make and use tools consistently, it is here that we could look for an explanation of the growth of the human brain.

As we saw in chapter one, it was Raymond Dart who first proposed the idea that hunting was the major impetus for human evolution, and in particular the evolution of the human brain. According to Dart, the 'killer ape' gave birth to the 'wise man', which is how we got our present name, *Homo sapiens*. William Calvin suggested his 'killer frisbee' hypothesis could account for growth in brain size. Remember, he believes that handaxes – the teardrop-shaped flint stone tools that were produced by *Homo erectus* all over Europe, Africa and Asia – were thrown like a discus at antelope and other herbivores on which they preyed while these animals gathered at water holes.

According to Calvin, firstly the hominids had to develop the complex skill of knapping handaxes. This involves advanced dexterity and motor skills as well as a basic, intuitive grasp of how flint fractures. Secondly, they had to develop the skill of throwing them accurately at distant and usually fast-moving targets. For that they would have needed greatly increased co-ordination, as well as an ability to pick up the skills of making and throwing.

While it's true that modern humans can throw objects with far greater power and accuracy than chimpanzees, our primate cousins occasionally manage to hurl a branch or two to scare off a competitor. But that's about it. Think about the huge reserves of cognitive power that would be needed to build a robot arm that could throw as efficiently as a human arm. The controls needed to manage movement, speed and trajectory are enormously complex. Then imagine trying to get the same robot to forge a handaxe from a block of flint, and we start to get an idea of the impressive cognitive complexity of early humans.

All of this adds up, says Calvin, to the need for a large amount of sheer neural computing power. But there's one very obvious problem with the theory. As we've seen, handaxes remained broadly the same shape and design for almost two million years. That's an extremely long time, considering the design lifespan of the average car or computer. But during that same time span, the era of *Homo erectus* and archaic *Homo sapiens*, brain size increased by an enormous 50 per cent.

Over two million years, through the lower and middle Pleistocene, there must have been a sweeping change in hominid lifestyle. Why else would our brains have swelled by this enormous amount? But during this period we did not invent any new tools; we appeared stuck with the traditional handaxe. There must have been another explanation.

The social brain

Anthropologists have long documented the tradition of passing on the expert knowledge and skills necessary to hunt, forage and gather, particularly in a harsh environment like the Kalahari Desert, home of the !Kung bushmen. During their teenage years, the !Kung become experienced trackers, collecting an impressive depth of knowledge on how to read the size, depth, direction and condition of animal tracks. They can tell how old the animal is, what sex it is, even whether it is lame or tired.

The ability to uncover that sort of information would surely have increased the average hominid's chances of survival, and this could have been a factor in the evolution of the bigger brain. Those hominids with the mental space to pick up new tricks and new knowledge would have found themselves at a strong advantage. This new knowledge does not necessarily become part of our shared psychological make-up, like a fear of snakes or a sense of direction. Learning the local knowledge of the landscape and the habits of animals is similar to learning to play the piano; it is not an adaptation, even though piano players do change the structure of their brains through years of practice.

But it does underpin one crucially important and significant part of what we consider to be human. We know that our human minds have a tremendous capacity to learn skills and pick up new knowledge, but it is what is at the heart of this learning process that counts. In order to learn, we must communicate with others, and the social nature of hominids and early humans was vitally important: how they lived as a group, found food as a group, protected and communicated with one another. This is undoubtedly a most enticing route to choose in the search of an explanation for the modern human brain.

Hunting is a prime example of this social imperative. Without the benefit of a high-powered rifle, hunting big game successfully not only relies upon physical prowess and individual skills, it also requires co-operation. It is difficult to catch large prey alone. Once prey is caught it makes sense to share the kill, otherwise much of the meat will go to waste. In large groups, sharing meat and other foods would have acted as a kind of insurance policy: if today you don't manage to catch anything, then tomorrow others will share their food with you.

And while living, hunting and gathering in groups would have meant everyone had a better chance of getting fed and therefore surviving, the process relied upon the hominids being able to communicate with one another. It implies the ability to predict how others will react, and to keep track of alliances and rivalries. Look at our own everyday behaviour. Though we cannot literally read other people's minds, we spend a great deal of our daily lives listening carefully to what people say, watching their faces, eyes and body language and trying to make sense of their behaviour. And nowadays, most of us are naturally skilled at it. The development of those skills, which some may regard as the very essence of being human and of higher thinking, must have been a real watershed in terms of the growth of brain size of our ancestors. Abilities like planning, memory, communication and consciousness of ourselves and others must surely constitute the twist in human evolution that marked us out from most other animals.

Bigger brains and higher intelligence have aided our survival so significantly in our evolutionary history that they've become something of a status symbol. They are our neurological version of the tail of a beautiful peacock; in the modern world our big brain has turned into something the opposite sex sometimes considers highly attractive, and the sexier we are considered by the opposite sex, the more likely we are to reproduce. Our intelligence lets us tell jokes, chat up potential mates, even write sonnets. But sex is far too important to our survival to rely purely on our having a way with words. In the next chapter, we shall take a look at how our instinct for sex goes much deeper.

chapter**three**

Sex and the Savannah

Sex and evolution

It has been said, with what I am sure is absolute accuracy, that most men think about sex every six minutes, while about 20 per cent of women think about sex at least once a day. Of all the human instincts we possess, sex shouts the loudest. We are obsessed, and even when our behaviour is not overtly sexual, we spend a great deal of time engaged in activities which are, in a fundamental way, connected with sex and reproduction: money, career, appearance, friendships, competition. All these aspects of human life are bound up in a complex web of sex-related impulses, whether we realize it or not.

Why is so much of our lives literally devoted to sex? Someone once said that sex is hereditary; if your parents never had it, chances are you won't either. Every single one of our ancestors had sex, and reproduced successfully – of that much we can be sure. It is a reasonable assumption that many of our ancestors' contemporaries failed to have sex. Perhaps they died before becoming sexually mature, or perhaps they had sex but failed to conceive. Perhaps they were last in the queue behind stronger, fitter, more desirable contemporaries and were denied access to the mating game. Evolution is extraordinarily single-minded when it comes to sex, because natural selection can only operate on the principle that those organisms which survive the period of development and maturation will go on to reproduce; those that reproduce are the 'winners' in the race of natural selection. Any animal that flunks the test is a *de facto* loser. When it comes down to the very nuts and bolts, life is a heady mix of surviving – and, yes, sex.

There is little doubt that it is remarkable that the human species is on the planet at all. One of the things that forcibly struck me on recent visits to Africa was the sparsity of early hominid remains. Each successive hominid species seems to have left only the tiniest fossil record, and each, of course (apart from *Homo sapiens* so far), died out. *Australopithecus ramidus* (five million years ago), *Australopithecus afarensis* (three million years ago), *Australopithecus robustus* (two million years ago), or *Homo habilis*, *erectus* and early *sapiens* must have been severely stretched just to stay alive on the threatening, hostile savannah. And there are suggestions that the founder members of our species who entered Europe were possibly just a dozen in number. The entire fossil collection we have from all hominids would fit into the back of one Land Rover. Possibly we were as rare as the modern blue whale, so the urge to reproduce must have been very strong for us to be here at all now.

So, if you've ever wondered why you have the instinct to have sex at all, it's because you personally are at one end of a very, very long line of sexual success stories. We are all designed to secure mates and produce children. As far as our genes are concerned, reproduction is the whole point. Of course, genes are not concerned about anything – they are not conscious entities – but it sometimes makes sense to think of them as having a point of view.

So far, every single member of the human race that has ever lived has been the product of sexual reproduction between a man and a woman. But possibly, in the not too distant future, children who have only one genetic parent may be born. The ability to manipulate the reproductive process, an inconceivable advance just one generation back, may allow us to bypass the biological tradition of the union of two half-complements of chromosomal DNA, one from the mother and one from the father. Instead, we may be able to implant one full complement of DNA from the nucleus of a normal, mature cell – possibly, say, a skin cell – into an ovum which has been emptied of its nucleus which contained the original maternal chromosomes. Through some clever chemical tinkering the egg may be fooled into thinking it has been fertilized, and it starts dividing. The embryo is therefore of just one genetic parent, and the child would be a clone – an identical genetic copy – of that single parent in the same way as identical twins are genetic copies of each other. Artificial cloning of human embryos has already been attempted, but so far none has survived more than a day or so. There is even doubt whether these embryos were truly 'alive' and normal. But it is likely the process will be refined, and the implantation of an embryonic clone into a woman's uterus, and its subsequent gestation, may well happen in the next few years, whether or not we try to outlaw the practice.

The juggernaut of advances in molecular genetics appears to be unstoppable. Still, in spite of all our deftness at intervening in the reproductive process we can safely assume ordinary, common-or-garden sex is here to stay. IVF and other artificial techniques are never likely to replace procreation in bed or on the hearth-rug because of our basic instincts. We prefer the old-fashioned method because we have an old-fashioned brain. Because evolution favours the ability to reproduce above all other considerations, sex is at the heart of what makes us human, at the very core of our make-up. As we shall see later on, many of our instincts, particularly those which have a bearing on our social behaviour – competition, co-operation, bringing up children, violence – are surrounded by competing urges, and often the strongest is the need to reproduce, to have sex, and to bring up our children in order that *they* can have sex.

Putting aside exceptions to the rule

Evolutionary psychologists intentionally view mankind as a single, homogenous mass. There is a good reason for this. Theirs is an attempt to find out what human minds have in common, whatever the race, geographical home or physical features of their owners. According to this view, any differences are surface wrinkles covering a common, stable psychological structure.

There is a sound evolutionary rationale behind this point of view. The most widely accepted theory of human evolution is that *Homo sapiens* evolved in Africa and only then spread out through Asia and Europe, and eventually the Americas and Australasia. For most of our evolutionary time, according to this theory (called the 'Out of Africa' theory), ancestors of every single member of the human race have been subject to similar evolutionary pressures on the savannah.

No-one would deny that our mental make-up differs substantially from one individual to another; some of us are greedy, some of us are violent, some of us are more interested in sex than others. And some people are interested in sex with others of the same sex. Across the world, an estimated 2 to 4 per cent of the population is gay. From the point of view of natural selection, homosexuality should be the ultimate dead-end. There can be no evolutionary advantage to being exclusively gay or lesbian. Yet humans are not the only species to indulge in homosexual behaviour: bonobos – pygmy chimpanzees – are often seen engaged in heavy petting with a same-sex partner. But it is an adjunct to the usual run of heterosexual mating, whereas humans can be exclusively homosexual, their only means of having biological children artificial intervention in the breeding process.

But are we really that surprised that humans are capable of behaviour that is entirely non-adaptive? A great number of human activities exist which run counter to our expected notions of natural selection: we engage, for instance, in self-destructive behaviour like drinking, smoking or driving around in metal boxes at high speed. Similarly, there are genetic conditions that cause potentially lethal diseases, yet they still exist in the population. Even genes that harm reproduction do not necessarily die out. My research group recently published evidence of a variant gene which results in embryos being unable to implant in the womb because the mucus produced by the womb is deficient in a protein called MUC-1. The result is sterility, and even IVF will not help.* So while one may think of humans as

* The deficiency in the MUC-1 gene may be treatable by artificially producing the normal protein and then placing it in the uterus at the time an embryo is also put there, where it would act as a kind of glue. What is interesting is that there are likely to be dozens of other gene defects which produce infertility by a broadly similar route.

machines designed to maximize their genetic fitness and reproduce as quickly and efficiently as possible, there is plenty of scope for variation in that task.

Homosexuality may be a cultural (or learned) phenomenon or it may be genetic – or, more likely, a mixture of the two – but it could be a perfectly 'natural' aspect of human behaviour while still remaining a minority activity in every generation. Perhaps much homosexual behaviour is simply sexuality at one end of the spectrum; perhaps, if it were not for the pressures modern society imposes both consciously and unconsciously on homosexuals, their behaviour might sometimes be quite different. Homosexual feelings, after all, are very common among adolescents who as they mature become increasingly inclined towards heterosexuality. Homosexuality may seem to go against the grain of evolutionary adaptation, but this does not mean that it is a 'bad' thing morally. Whatever the basis of homosexual behaviour it still makes sense to explain sexual instincts on the basis of more typical behaviour.

When we're thinking about the nature of human psychology, it helps to take into account the fact that, broadly speaking, we all have the same mental architecture. Our minds are those of our savannah ancestors, all of whom evolved into *Homo sapiens* under the same environmental conditions and the same evolutionary pressures. That's why, for our purposes, we're more interested in the rule rather than the exceptions to the rule. The question of why some of us turned out like this rather than that, why people have a particular psychological preference or an unusual set of instincts, is important, interesting and illuminating, but it is not what this book is about. For the purposes of shining light on our human instincts, those the majority of people share, variations on the theme have to be put aside.

Evolutionary psychologists quickly found that they had to be careful. They are extremely sensitive to accusations that their work gives succour to racist ideologies. Previous generations saw genetic theories of human behaviour used to promote eugenics, to contribute to a pseudo-intellectual foundation for Nazism. But modern practitioners are careful to point out that differences in skin colour, hair type and facial characteristics, even predisposition to certain diseases, are mere superficialities, and on that point I am sure they are right. There appear to be very few important physical differences between races, and no mental differences that can be attributed to different racial or genetic types. In fact, genetically speaking, there is no more major difference between my DNA and that of an African bushman than there is between yours and your next-door neighbour's.

There is, however, one very fundamental split in the human family. It is said that the contrast between male and female psychology is the only difference that really makes a difference. The chasm of psychology and sexuality which divides the two genders runs wide and deep.

A wide-reaching study of men's and women's sexual attitudes, called *Sex in America: A Definitive Survey*, was published in 1994. Faced with abundant evidence of significant differences between the genders, the authors stated there was 'no reason to believe these differences . . . reflect some sort of genetic imperative'. Researchers with a taste for evolutionary explanations took the bait and began to map out a genetic foundation of human sexuality.

Man seeks woman

Homo sapiens *male seeks female of the same species with maximum future reproductive value. Must be young, with clear skin, symmetrical facial features, a waist-to-hip ratio of 0:7, lustrous hair and preferably an absence of sores and lesions.*

Sex begins with looking for a mate. Most *Homo sapiens* have a complex series of tests their potential mate must pass if they are to prove suitable. These, it should be said, are idealized features of a mate, and ideals are, of course, toned down with experience. But for the majority of human males across the world there is a surprising consensus on what constitutes a sexually attractive female.

'Future reproductive value' is an elaborate way of describing how many children a particular woman can bear in the future. A woman who is young, fertile and healthy has a better chance of bearing a number of children who themselves will be successful and will go on to reproduce. Infant and child mortality must have been high on the savannah. The most dangerous journey humans take is one of just four inches. That journey down the birth canal, so easily obstructed because of the tight fit of a human cranium in the maternal pelvis, with no medical assistance at hand must have resulted (as it still does in many primitive societies) in high mortality or morbidity among newborn infants. And disease, drought, famine and predators must have frequently posed grave threats to a vulnerable infant. In genetic terms, then, it was vitally important for the male to make sure his mate had the ability to produce plenty of babies so that at least one or two of them survived to adulthood.

Perhaps as a consequence, men in all cultures tend to find younger women more sexually attractive. The Yanomamo are a case in point. They are an indigenous people living in the northern Amazon forest in Brazil and Venezuela. Living in almost total seclusion, this essentially pre-Columbian society numbering some twenty-three thousand people in all are one of the last human groups to have remained largely untouched by modern human 'civilization'. They are of great interest to anthropologists because they tell us something about the nature of man 'in the wild'. Sadly, with the increasing destruction of the rainforest these remarkable people are greatly threatened.

A Yanomamo tribesman would say he prefers a 'moko dude', which normally means a harvestable fruit, and in this context means a young, sexually mature woman with no children. This human desire is tempered by the realities of sexual competition, and also by the fact that potential mates are often taken from among our own peer group, but the actual age difference between married couples across a number of different countries averages out at almost three years, with the man being the older partner. This is particularly interesting, given that males are more susceptible to death by disease and trauma, and throughout the world tend to die younger than females. But, of course, unlike men who can occasionally carry on impregnating their partners when over eighty years old,* the human female starts to enter her menopause some time after the age of forty, when her fertility drops precipitately.

Selection over hundreds of thousands of years will presumably have favoured those men who seek out healthy, fertile women. Unfortunately, men cannot tell how fertile a woman is just by looking at her; instead, they have to rely on other indications. Clear skin, shiny hair and full lips may seem like reasonable indicators of general health, but they are not particularly good indicators of fertility. But youth, undoubtedly, is a generally reliable one, and humans, both men and women, are remarkably good at assessing a person's age simply by glancing at them.

Of course, though, looks count. Males are more attracted to symmetrical faces, and can easily spot even tiny asymmetries in a face. This is linked to our inborn ability to easily recognize different faces, and men apparently redirect this talent later in life towards selecting desirable mates. In fact, there is a fairly precise formula for the structure of the perfect female face, drawn up by the ancient Greek philosopher and mathematician Pythagoras. He held that for someone to be considered 'beautiful', the ratio of the width of the mouth to the width of the nose should be 1.618 to 1. This figure should also hold for the ratio of the width of the mouth to the width of the cheekbones. If the face of your favourite supermodel or Hollywood starlet is measured, there's a pretty good chance they'll match up to Pythagoras's formula.

Does our culture teach us what we regard as the accepted characteristics of what we call beauty? Nowadays, we are bombarded daily with images of supermodels and Hollywood starlets who must have the right look to 'make it' in their profession. But in a study carried out at the University of Texas, a group of infants aged just three months old were shown a series of photographs of human female faces and the babies' responses to each face were measured. All the infants responded much more significantly to faces which

* Pablo Picasso was well into his seventies when his last child was born, and Havelock Ellis, the famous sexologist who wrote *Physiology of Sex*, reports one man who produced a child at the age of 103.

meet the culturally accepted standards of beauty. Clearly, by testing three-month-olds the researchers could be sure that their subjects hadn't been watching the Oscars or reading *Vogue*. This study suggests we are born with instinctive concepts of beauty undoubtedly reinforced by cultural learning as we grow up.

During my work on the TV series on instinct, I had the pleasure of meeting Professor Peter Hammond at the Eastman Dental Institute in London. He scans human faces using a form of computed photographic modelling and can compare human faces in health and disease, and during ageing. His powerful computer is able to assess the shape and dimensions of each human face it scans and can morph changes induced by different genetic characteristics. I asked Professor Hammond to show me an average human face at different ages. His computer rapidly assessed over two hundred faces on magnetic storage and produced an average six-year-old and an average adolescent. What I found so engaging was that these faces, with their 'perfect' ratios, looked eerily like cherubs painted by Renaissance artists such as Bellini or Titian.

Putting the face to one side, what about the female form itself? As we know from popular culture and fashion, men find the ratio of a woman's hip measurement to her waist measurement an attractive feature. And as it turns out, the hourglass figure does indeed have some evolutionary significance.

It is true that different men and different cultures are attracted to varying types of female body, but there is one apparent universal attraction. Men the world over choose mates whose hip measurements are much bigger than their waistlines. The preferred waist-to-hip ratio appears to remain the same, at 0.7. This ratio remains constant for Marilyn Monroe and Sophia Loren as well as Audrey Hepburn and Kate Moss. It even holds for ancient twenty-five-thousand-year-old Venus figurines, the small stone sculptures of women found across Europe and Asia, possibly connected to fertility cults. Although many of these figurines tend be distorted by being enormously fat, all of them adhere to the 'golden' ratio.

Possibly, the waist-to-hip ratio may tell a man something about a woman's health. After puberty, oestrogen causes increasing amounts of adipose fat to be laid down on women's thighs, buttocks and breasts – fat which once was likely to be important for survival for both mother and child in conditions of scarcity. Even today, once a woman's fat deposits drop below a certain percentage of her body weight, she will stop ovulating. The classical example is the condition anorexia nervosa. Young women who, for rather unclear psychological reasons, starve themselves become excessively thin. Once their body weight drops below a certain level, ovulation ceases, and even menstruation stops eventually. Pregnancy under these circumstances is normally impossible, and what is interesting is that usually no amount of ovulation-inducing drugs will be successful. And if an unlikely pregnancy does occur under these circumstances, the loss rate by miscarriage is very high indeed. Even milder, non-pathological loss of weight can cause severe fertility disturbance. Female athletes, ballerinas, models and body builders are prone to menstrual disorders and are frequently infertile. Just possibly, then, if a man can see that a woman's hips are larger than her waist, he may be more sure that her body is healthy, functioning and, most important of all, fertile. Certainly, this was likely to be true under the conditions which must have been common on the savannah. If men think that women with hourglass figures are sexy, they'll probably end up choosing one as a mate. This would increase the chance of fathering lots of healthy, robust offspring to inherit a penchant for small-waisted women. Just repeat this for thousands of years and you may see why most men on the planet get a kick out of curves.

But contrary to popular belief, men are not solely interested in women's faces and figures. I'm not talking about personality, either. Men also look for behavioural hints, for example a woman's posture, or how vivacious or sprightly she appears. In evolutionary terms, it's all about the physical capacity to reproduce successfully, but a desirable female mate, in addition to possessing these physical characteristics, should have a certain reputation – or, rather, lack of one. Why do the Yanomamo place such store in an apparently chaste woman with no children? Because a chaste woman is more likely to be a virgin and less likely to have engaged in prolific sexual activity.

Virginity has been highly prized in many cultures. An advert appearing a few years ago in an African newspaper is indicative of this preference: 'European bachelor 30 years, salary 200LS, wishes to Marry Educated Cultured Virgin, nationality no consideration.' Why this obsession with sexual inexperience and chastity? The worst situation, evolutionarily speaking, in which a male can find himself is unknowingly bringing up someone else's children. The evolutionary consequence for the male if this happens is disastrous. He would expend time, energy and resources caring for a child who does not carry his genes. Conversely, a competitor who impregnates another's mate has all the benefits of passing on his genetic inheritance with none of the drawbacks of having to provide for the baby and its mother. So the safest bet for a male worried about his partner's infidelity is to choose a woman who appears to be chaste. Being sure of one's paternity is of overriding importance. Perhaps, in consequence, sexual jealousy is a fundamental feature of human sexuality.

The desire for chastity sits oddly with the fact that men are almost always far more keen to have sex with a potential mate than are women. Robert Wright points out in *The Moral Animal* that the more promiscuous a woman, the less likely she is to be a faithful long-term partner – at least in the eyes of the jealous male. Men seduce; they cajole a woman into bed, perhaps knowing that if she accedes – at least, too soon – then she will be labelled an unsafe bet for the future. This is what has become known as the Madonna-whore dichotomy. There is a constant tension between the desire for sex and the need for a faithful mate. Wright suggests men have evolved to set a kind of 'test' for potential partners, a test which involves persuading them to have sex, then checking to see if they have self-restraint.

The value men place on youth and chastity is an instinct which seems to be reflected in the more extreme corners of society as a sexual obsession with young girls, leading to horrifying sexual abuse. The internet has proved to be a breeding ground for pornography, with an expanding market for paedophilia involving young girls, boys and sometimes even infants. The underground groups that have been investigated and exposed to date underline the extent of this problem and the lengths to which their members will go to find and abuse children. The 'cult' of the virgin is also reflected in some dangerous modern trends. In South Africa, which has one of the worst HIV infection rates in the world, there is a belief that sex with a virgin can cure AIDS. Witch doctors in the country have been accused of spreading this notion, and there has been an appalling rise in the rape of young girls, even babies. It seems that even the most deeply rooted of human sexual instincts can have terrible consequences.

On a day-to-day level, millions of years of evolution do not guarantee our sex lives will be harmonious or that our sexual preferences will make a perfect fit with modern morality. The 'shopping list' of desirable female characteristics appears to fit neatly with the old stereotype of selfish, libidinous, male lust. Surely the modern sensitive male chooses a partner on more rational, cerebral grounds? What of the importance of sharing the same sense of humour, the same perspective on life, the same political or moral convictions? What about an emotional or spiritual 'connection'? Physical attractiveness is definitely not the whole story. We do look for more than that – indeed, we look for love, friendship, intelligence and a spiritual connection – but these and similar ingredients of mate selection transcend our most fundamental sexual instincts.

There is, however, one very clear point to be made. The sexual engines of men and women are running on entirely different kinds of fuel. David Buss, an American evolutionary psychologist, has conducted pioneering studies on the sexual preferences of men and women. His work suggests that, overall, both men and women have rated physical good looks as increasingly important over the past fifty years, coinciding with the carpet-bombing of modern culture with images of attractive people. The everyday sight of the Wonderbra girl or a dishy Hollywood action hero has raised our standards, which, given the lack of supermodels and movie stars on the dating circuit, can be something of a psychological catch. But the basic difference between men and women remains: men rate physical attractiveness much more highly than women do.

What women want

Female of child-bearing age seeks older male, high income, risk-taking altruist, but dependable and faithful. Square jaw and symmetry desirable, but not essential. Untrustworthy men with a moustache need not apply.

Before female readers get a little irritated by men's evolutionary predilections for waist-to-hip ratios and symmetrical faces, consider an interesting study investigating personal ads – placed by men – which were most successful in eliciting replies from women. The following factors were found to be the most important in stimulating female interest: age, level of education and income. All three are predictors of financial stability. Time and time again these, rather than a 'nice personality', stand out as the defining features of what women want in a man.

Needless to say, women by no means ignore a man's looks entirely. The traditional visual stereotype of a healthy, testosterone-rich male, tall, square-jawed and muscular, is certainly sought after. Just look at the cover of a typical Mills & Boon novel. And through the ages, men's dress has tended to accentuate these traits: the square, boxy shoulders of a suit, or the moulded, muscular curves of Roman armour, which flatter and augment the physique of the wearer. But size and shape are definitely secondary considerations. Women do seem to have evolved a somewhat more sophisticated and long-term measure of what makes a good mate. They are much more focused on the longer term – particularly, according to the studies, the financial future – rather than the size of a man's biceps. Yet again we seem to have unearthed another sexual stereotype, this time of the status-obsessed, gold-digging woman (who often prefers an older man).

In David Buss's survey into mating preferences, which covered thirty-three countries and questioned over ten thousand people, women on average placed twice as much value as men on their potential mate having good financial prospects. Some cultures valued it more than others: Nigerian and Japanese women, for example, rated financial prospects higher than women from Finland and South Africa. But the female interest in the finances of their potential mate was always higher than their male countrymen's.

Women also preferred their men to be older (which, luckily, corresponds with men's desire for younger women), but to a large extent this can also correspond with financial security. Older men are paid more, they tend to be more careful with money and they tend to be more secure and settled.

Women, however, are not solely interested in settling down with a well-paid accountant, or indeed, at a stretch, a tenured university professor. According to a recent study by Robin Dunbar at the University of Liverpool, they like men who take risks, especially those who take risks on behalf of other people. *ER*'s Dr Ross, played by George Clooney, is a perfect role model: he was heroic and performed regular acts of bravery on behalf of others. Even better, he was a paediatrician – kind to young children, too (the scriptwriters really did go to town). In Dunbar's study, women were given a selection of hypothetical men, from a fireman who had won a medal for bravery to a golf-playing supermarket manager. Firemen, who display their altruism and bravery on a regular basis, won the day.

The typical heroes of a Mills & Boon romance possess all these qualities. They are always strong, risk-taking and brave. The only roles available for weak, sensitive cowards are as foils for the more desirable hero. Dunbar suggests that women interpret this heroism as an ability to protect them and their offspring. In a world where danger was lurking around every corner, cave and hillock in the form of sabre-tooth tigers as well as predatory males from rival bands of early humans, bravery was a cardinal quality much in demand in a mate. But Dunbar found that the female long-term mating strategy would avoid those males who take selfish or unnecessary risks. Roguish, promiscuous men were considered as candidates for a one-night stand (perhaps indicating a female tendency to dip into the gene pool surreptitiously), but they did not score well for long-term relationships or marriage.

Erotic fiction and pornography also offer a somewhat illuminating window on what women may want from a man. In recent years, there has been an emergence and rise of pornography aimed specifically at women, and erotic novels written by women for women. While men may sometimes feature purely for sexual pleasure, this sort of erotica differs quite markedly from the male variety. Here, firstly, men are interested in women's sexual pleasure and, most importantly, in taking part in loving and emotionally charged relationships, even if they are only short term. There's by no means the same level of physical explicitness as in male pornography. The emphasis is on the idea of sexual liaisons between loving couples – emotionally torrid encounters, not just emotionless one-night stands.

The latter sets the tone of most pornography aimed at men. Men, famously, are more easily aroused by the visual image, even if it is a poorly reproduced copy of a drawing of a naked woman or even a very pixellated image on a computer screen. When it comes to stories, men like a very different kind of narrative. They are often drawn to stories that do not end with a loving and long-lasting relationship; rather, they prefer descriptions of sexual acts with anonymous strangers, none of whom is likely to stick around for breakfast. Their 'pornotopia' is generally dominated by the free-and-easy, sexually available, no-strings-attached nymphomaniac.

The gap between men's and women's sexual psychology is wide. Women are interested in character, commitment and security; men are interested in physical female attributes and anonymous sex. Does the mismatch mean that neither sex is altogether content with what they get? The situation is not entirely hopeless. These are tremendously powerful instincts, but they can play a greater or lesser role in our sexual identity. For some people, choosing a mate is not dependent on facial symmetry, a bank balance or an ability to perform open-heart massage without breaking into a sweat. Some women are distinctly not interested in commitment and seek out casual sex. Some men do not give a hoot for waist-to-hip ratios, symmetry and clear skin, and instead look for a Capricorn with an addiction to watersports. But they are untypical.

Eggs v. sperm

How do these appetites for the opposite sex translate into the real world? Ideally, our genetic programme would guide us efficiently towards the best possible evolutionary strategy. We would lead uncomplicated, equitable and happy sex lives. But, the world over, humans are all too frequently filled with conflicting desires, unconscious urges and a psychological agenda that serve to inflame the battle of the sexes. For modern *Homo sapiens*, mating is a tiny bit like playing chess while your opponent holds a gun to your head. And psychoanalysts everywhere would bet their BMWs that sex is not going to get any easier.

One of the largest cells in the human body is the egg, or ovum. At the rate of one a month, over the course of around thirty or so fertile years, a woman may release just four hundred ripe ova from puberty to menopause. It is true that she will have far more eggs in her ovary (around three hundred thousand at puberty) than she will ever ovulate, but these eggs simply die in the ovary one by one, with ageing. By the time of the menopause there will be virtually no eggs left. Assuming one of the four hundred eggs she ovulates is fertilized, the baby she bears will take nine months to gestate, which represents a huge biological investment. If we take into account the time taken to raise the child, breast-feed it and attend to its every need until it slowly gains self-sufficiency, then it becomes clear why women must be extremely careful as to who provides the other half of the genetic package for any particular child. If she makes a mistake and becomes pregnant by an unsuitable mate and father who does not protect his family, she will be in trouble. She will have wasted massive resources. Such a waste would have been even more critical on the savannah, where a high infant mortality would reduce the chance of genetic continuity.

Compare this with the investment made by the father. An adult human male may produce around a hundred million sperm every day. He could father hundreds, if not thousands, of babies if he was given the opportunity. A sperm is the smallest cell in the human body, and its production involves the tiniest expenditure of energy. His role in any given conception simply involves a few minutes in the sack, ejaculation and maybe a cigarette later.

According to the *Guinness Book of Records*, the Emperor of Morocco, Moulay Ismail the Bloodthirsty (1672–1727), holds the record for fathering the largest number of children – apparently (but exactly who was counting them?) a staggering 888. On the other hand, the most fertile woman in recorded history, a Russian peasant by the name of Madame Vassilyev, gave birth twenty-seven times, producing a total of sixty-nine children: sixteen pairs of twins, seven triplets and four quadruplets between, it is said, 1725 and 1765. Her extraordinary number of offspring still only totalled well under 10 per cent of the number fathered by the rather impressive Emperor Moulay the Bloodthirsty. So it is likely that the very nature of our sperm or eggs is to some extent driving our sexual behaviour in seeking out a mate. From the male perspective, it's evolutionarily desirable to impregnate as many females as possible. From the female perspective, it pays to be cautious and to choose a mate with great care.

Consider the rather unpleasant female Black Widow spider. The high value of the female's egg becomes especially clear in her everyday sexual behaviour (or perhaps that should be misbehaviour). One species of Black Widow, the Australian redback spider, is particularly romantic. The male and the much larger female begin with a four-hour ritual of courtship. But it's not entirely red roses and serenades. Once they begin to mate, the male uses one of his reproductive organs to somersault himself backwards into the mouth of the female; they continue mating as the female begins to bite chunks out of him. As she munches away, the male is able to continue copulating and transferring sperm. Often he survives the first attempt, but if he does he returns immediately to court the female. This time she finishes him off, devouring every last leg.

Sexual cannibalism is unknown among other animals, but Black Widows have evolved this extraordinary behaviour for very good reasons. They live in an environment where their staple food, cockroaches and beetles, is often in short supply. The females live for up to eighteen months and have just one mating season during that time, in the spring, so they must take full advantage (*carpe diem* should be their motto). Mating without enough food for both mother and babies would be reckless to completely pointless, for both parents. Their solution is for the male to sacrifice himself in order to provide the nutrients for both mother and babies. Better to mate once successfully and die than never to mate successfully at all.

Tactics change, but the primary consideration, even to the point of eating your partner, is to protect the female investment in the egg. We have the same long-term aims, but men and women have different ways of achieving them. Men strive to sleep around, or at least they may be tempted to sleep around if given the opportunity. In one entertaining experiment, male and female researchers accosted students of the opposite sex on a university campus and asked them if they wanted to have sex. Some three-quarters of the men said they would agree to immediate sex; none of the women agreed (although a small minority accepted a dinner invitation).

In a long-term relationship, a sense of loyalty, the mores and morals of any particular culture and the unwritten rules of the institution of marriage often dissuade the man from attempting to sleep with other women. But if he refuses an opportunity for extra-marital sex, it is as much a case of overcoming physical temptation as it is a moral choice. Women, on the other hand, generally have less interest in sleeping with a complete stranger. Despite modern contraceptive techniques, which provide a virtually cast-iron guarantee that she will not conceive (at least in comparison to riskier systems like the rhythm method), women are not built to seek out sex with a large number of men, strangers or otherwise. They are much more interested in quality rather than quantity. They have relatively few eggs, so they've got to be choosy about who gets to fertilize them.

The whiff of love

The notion of chivalry and courtly love goes back to medieval times. We humans are still firmly attached to the notion of romance. It is considered to be a force that, though irrational and sometimes inexplicable, is above all biological or genetic considerations and has nothing to do with evolutionary strategy or natural selection on the savannah. We would like to think that there is always room for a spiritual, indefinable idea of love, but biologists are chipping away at the walls we have erected between biology and spirituality, inserting their biological explanations in an edifice we always thought was reserved for human motivation, reason and emotion. Might love be another casualty?

In 1976, a group of researchers in New York began investigating the genetics of mating, and they started by looking at laboratory mice. They concentrated on an important group of genes called the MHC genes, or the major histocompatibility complex. These genes are present in nearly all the cells of mammals and they play a major role in the immune system. The MHC genes produce proteins which spend their whole lives trying to define 'us' (in immunological terms, that is). By being able to recognize us and our cells, they can then recognize foreign bodies or pathogens such as invading micro-organisms which could potentially cause disease, then send out the signal to mobilize the body's biochemical defences. They are critical in transplantation and are responsible for the rejection of 'foreign' organs. Like most genes in the human genome, MHC genes vary from individual to individual. But MHC genes vary considerably, which is why so few of us – except, for example, identical twins – have identical tissue types. It is this phenomenon which makes it so difficult to find compatible bone marrow donors for the treatment of such diseases as leukaemia, and why we need, almost invariably, to suppress the immune system with drugs after a kidney transplant, even at the risk of causing cancer, diabetes or high blood pressure.

If two mice from an average colony mate, some of these genes will be similar while others will not. But remarkably, the US researchers found that mice were more likely to mate with partners who had *dissimilar* MHCs. The experiment was repeated for other mice who were genetically more diverse and had been raised in an outdoor environment, which allowed them much more of a choice of mate. The results were the same. They all seemed to prefer mates with dissimilar MHC genes; opposites really did attract. These mice appeared to have evolved a mechanism to 'sniff out' a certain type of biological mate.

The Hutterian Brethren, or Hutterites, are among the most studied humans on the planet. These people, who now number some thirty-five thousand, are descended from members of a strict Anabaptist sect whose origins are to be found in sixteenth-century Switzerland. Because they were so persecuted in Europe, in the 1870s most emigrated to America, to South Dakota, Montana and western Canada in the main, where they now live. They are relatively isolated and still speak Tyrolean German. They live in small colonies on strict socialist principles (none own property) and are devout Christians. None of the Hutterites marry out of the community, and all practise strict lifelong monogamy and avoid the use of contraception.

Some years ago, Dr Carole Ober, a geneticist at the University of Chicago, decided to take the mice compatibility studies a stage further and began studying MHC genes in the Hutterites. The Brethren, who have been helpful to biological research in all sorts of ways, were a perfect group for investigation of the mating choices of a contained group of people (in this respect their situation mimics the situation of early humans, who would have lived in groups of not much more than one hundred).

Despite their choice of marriage partner being restricted to other Hutterites, Ober's study revealed that these people did appear to exercise a different type of choice of partner: their choices seemed to be influenced by their partner's MHC complex. Like the mice, Hutterites tended to choose mates whose MHC was different from their own. But why would the make-up of the MHC be such an important consideration in the choice of a mate? What on earth could possibly be so sexy about an immunological mechanism?

It turns out that there could be something very attractive – in evolutionary terms, that is – to choosing a mate with a different MHC. There is a payoff in having a mate who is genetically dissimilar to oneself. We all carry defects in our DNA. Every reader of this book will carry some genetic defect which could be fatal in his or her child, but providing we do not mate with someone who has the identical defect in their DNA at the same point on a particular chromosome, our children will nearly always be protected. The normal DNA sequence from the defect-free parent compensates. And the MHC gene complex may be an ideal marker for genetic similarity or dissimilarity.

Inbreeding is dangerous. Inbred humans will always have a higher incidence of genetic disease. When I worked as an obstetric registrar in training in one particular part of rural southern England where people seldom travelled far from their birthplace and marriage with relatives such as cousins was common (I had better not identify the location much more precisely), I delivered an unusually high number of abnormal babies. I also saw a number of women who experienced unusually frequent miscarriages. Implantation of the embryo is more successful when the fetus is genetically dissimilar from the mother, and the higher incidence of miscarriage in women who have conceived with a male who is genetically related may be a protective mechanism which promotes healthy genetic diversity in the population. In the central part of Leviticus there is a series of strict instructions about forbidden marriages. But these interdictions are not just biblical. It is of considerable interest that incest has been a taboo in practically every known human culture, and perhaps almost certainly for this reason. The dangers of inbreeding are too much to bear, and individuals who practised it have tended to die out.

But how do we detect if a potential mate has similar MHC genes to our own? One answer is provided by studies involving T-shirt sniffing. For example, Claus Wedekind and colleagues at the University of Berne tested and typed the MHC genes of a number of female students. Then they asked a group of male students, whose MHC genes were also typed, to wear cotton T-shirts so that their body odour permeated the fabric. The T-shirts were taken to a laboratory where they were sniffed by each of the women in the sample. The women rated them according to how 'pleasant' they found the smell. They preferred the smell of the T-shirts worn by men with dissimilar MHC genes to their own.

During making the television series, I did a similar test. We were greatly helped by researchers in the Department of Psychology at Newcastle University. The delightful Dr Craig Roberts of the Evolution and Behaviour Research Group had arranged for six female undergraduates, whose tissue type had been previously determined and whom I had not met, to sleep for two nights in identical white T-shirts. They were not allowed to use perfume or deodorants, and had avoided smoking and strong-smelling foods such as curry or onions before the experiment. After two nights of nocturnal wear they sealed the T-shirts into plastic bags and brought them into the laboratory. There, Craig placed them in identical plastic jars covered at the mouth with tin foil. Once the jars were on the bench, Craig asked me to cut a small hole in the tin foil and to sniff each jar for a minute. I was then asked to place the bottles in order on the bench, with my most preferred smell on the left and the least preferred on the right.

What was interesting to me was that the differences were genuinely tiny. They all smelled vaguely sweaty, and all mildly unattractive. Some seemed a bit less attractive than others, but really only marginally. But I was genuinely surprised to see, when I stripped the covers off the labels on the bottles, that my ranking order was closely linked to the correlation between each girl's tissue type and my own (see photograph). Considering how sceptical I had been before the experiment, and remembering that there was no way I could wittingly have influenced the outcome, this seems genuine proof of the extraordinary power of pheromones.

Pheromones – subtle odours emitted by each of us – may well influence our choice of mate. The results of these experiments do suggest we can literally sniff out suitable mates. But how is a complex piece of genetic coding translated into body odour? This is not clear, but it is very likely to be due to the bacteria each of us carries on our skin all the time. Bacteria produce and change body odour, and the precise smell depends in part on the type of bacteria. It may well be that particular bacteria are favoured by one person's immune system rather than another, giving us each a slightly different olfactory signature. It is quite interesting that groups of women who are taking the contraceptive pill may perceive smell differently, and may themselves smell different. Could this mean that women who meet and forge a long-term relationship with a partner while they're on the pill are likely to have chosen a different mate had they been pill-free? It's a contentious idea, and the jury's still out.

It seems that our romantic notion of eyes meeting across a crowded room is, in part, rather a case of a refined olfactory ability to pick out a biologically suitable partner. Another broken string in Cupid's bow – our dreams of romantic love seem to be crumbling. John Donne wrote:

I am two fools, I know
For loving, and for saying so
In whining poetry.

He certainly would have struggled to find the words to convey such a mechanical form of passion, but the phrase 'animal attraction' seems rather apt. Poetry may suffer, but there'll probably be an entrepreneur or two attempting to take advantage of these discoveries in the future. Perhaps a personalized artificial scent based on the DNA of the person you plan to seduce? A lock of hair for analysis and a credit card and you'll be all set to woo to your heart's content.

The chemistry of infatuation

When two people do finally sniff each other out and fall in love, there is a period of time – on average lasting for around eighteen months to three years – during which passion is at its height. It is a time of exhilaration, of being 'high' in a natural and pleasure-filled state of excitement.

This state of mind has a lot to do with a love-drug called PEA, or phenylethylamine. The chemical is produced in the brain in large quantities during this fiery period of ardour and *amour*, and its effects are somewhat similar to amphetamines, or speed. PEA collects at the nerve endings and helps the electrical impulses jump across the gap, or synapse, from one nerve cell to another. When nerve cells in your limbic system are fuelled by PEA, you'll feel high, full of energy and sometimes euphoric.

PEA is not just present in the limbic systems of love-struck couples, it accompanies other intense experiences too. Parachute jumpers' PEA production goes into overdrive during free-fall, an experience they describe as accompanied by a feeling of great exhilaration. It is this buzz which some people probably get from bungee-jumping; very commonly, people who have just jumped experience a period of heady elation for some time afterwards. It may be that the emotional high, whether it's caused by love or throwing yourself from on high, itself stimulates PEA production, rather than the other way around. But what is clear is that either we fall in love to get our PEA injection, like an addict finding his next fix, or the PEA is our 'reward' for falling in love (and if you're unlucky in love and down in the dumps, as women the world over know, you can always get a fleeting PEA-fix by eating a bar of chocolate).

The evolutionary psychologist Helen Fisher has argued that our culture probably determines whom we love and when and where we love them, but it's undoubtedly the effects of PEA saturation in the brain which direct how we feel as we get pulled into the love affair. That chemical aspect of infatuation must have evolved over thousands of years to bond us to the person in question.

Whichever comes first, PEA or love itself, the trip, like all good things, must come to an end. Love literally is a drug, and a highly addictive one at that, and like all addictions there is a law of diminishing returns. The positive effects wear off after a certain period of time; the nerve cells of our brains become tolerant to the unusually high levels of PEA, and we come back down to earth. Once the honeymoon's over, passionate romances tend to mellow into what Fisher calls 'attachments'. In this phase of a relationship, rather than producing PEA your brain starts to pump out endorphins, brain opiates that are more like morphine than speed and serve to calm the mind, kill pain and reduce anxiety. From that point on the survival of our relationships depends on other feelings and desires: sexual attraction, of course, friendship and interdependence. But as we'll see, those psychological bonds can be extremely fragile.

Suspicious minds

A clever entrepreneur with a good understanding of the basic psychological drives of his customers recently started selling 'infidelity kits'. These are essentially DIY sperm detectors for suspicious lovers. In fact, you can buy them over the internet as 'CheckMate', 'The Original 5 Minute Infidelity Test Kit'. After covertly getting a few moments on your own with your partner's underwear, the kit can be used to swab the panties for traces of seminal fluid. It's the modern equivalent of lipstick on the collar. And the test goes further than that: for some two hundred dollars (plus an extra ten dollars if you want the garment returned) you can send a pair of knickers to the laboratory whose technicians will test them for DNA. For another five hundred dollars the lab will make DNA comparisons between the incriminating stains and a seminal fluid sample taken from an unsuspecting partner.

It turns out that sexual jealousy is an impulse we've almost certainly inherited from our ancestors on the savannah. A less suspicious species than ours would have become extinct many years ago. Early humans certainly didn't have the benefit of sperm-testing technology, but they still knew that sexual infidelity was a threat, especially to the male partner. From an evolutionary standpoint, bringing up another man's child is, as has already been stated, a grievous misuse of resources; the whole point of monogamy, to be harshly utilitarian, is that a partnership provides protection and resources for your own biological children. When it comes to our leaving our genetic inheritance to the world, we are blessed with tunnel vision, and anything which interferes with the process is a grave threat. It's not surprising that the danger and fear of cuckoldry casts a dark shadow across much of men's sexual psychology. The cuckolded male is depositing energy, time and resources into someone else's genetic bank account; somewhere deep in the male psyche is a voice which says *Avoid this scenario at all costs!*

There's one huge stumbling block for men, though. *Homo sapiens* is unusual in that a woman's monthly period of fertility is hidden, unlike species such as baboons or bonobos (pygmy chimpanzees) who are not shy about broadcasting their fertile condition to the entire troop: during their period of 'oestrus' their entire genital area swells and turns a bright shade of pink. Human females do no such thing. They themselves often tend to be unaware when they are ovulating. This modesty has two major consequences. The first is that humans desire and are able to have sex on any day of any month, whether or not a woman is fertile. We are extremely unusual in this respect, since most other species save their sexual energy for the time when it can be used most profitably. The second is that in a monogamous coupling, men have constantly to be on their guard because female infidelity is potentially dangerous at any time. They can never relax, and what's more, they can never leave their mates to their own devices.

Having said all that, it is certainly true that some women do feel rather more sexy around the middle of their cycle. Although this may be the result of the ovary producing its peak levels of the female hormone oestrogen at this time, women in general do not feel any more aroused if they are taking this hormone as a drug, for example in the contraceptive pill. To me, this suggests a basic instinctual mechanism, an instinct which possibly has been inherited from those animals who experience an oestrous cycle.

Concealed ovulation is a clever ruse. It goes hand in hand with internal fertilization. In some species, like the salmon, fertilization occurs outside the mother's body, which leaves a male in little doubt as to the paternity of his offspring. But in all mammals fertilization occurs internally, within the female's body, so while all women can be sure that they are the true genetic mothers of their children, men face a constant uncertainty. One African culture underlines the biological truth with the saying 'Mama's baby, papa's maybe'. Just possibly, both concealed ovulation and internal fertilization evolved as mechanisms to ensure that a woman's mate was attentive all month long. They reduce the risk of desertion by the male, which itself reduces the risk of the male forging relationships with other women. Were these, perhaps, the very beginnings of a trend towards monogamy in human culture? They were also the spark for our capacity for jealousy.

Jealousy rests on a hair trigger. It can arise from the most innocuous of circumstances and be fired by the faintest hint of infidelity. Othello's rage can be awoken from 'trifles light as air'. Jealousy is literally a force of nature, and Shakespeare knew that it makes for the best and most tragic of dramas.

The horror of female infidelity can provoke social ostracism. Men who are unfortunate enough to be cuckolded can endure humiliation at the hands of their male peer group. A Greek man whose wife has been unfaithful is called a 'Keratas',* or cuckold – a painful slur on one's character, implying weakness or impotence. Italian-American men have been known to go to the expense of returning

* Those in search of a good meal cooked in olive oil could do worse than visit the Restaurant Mylos, in Platanias, Crete. The building is decorated with numerous animal horns which gives it its local name, 'Keratas'.

to Italy to try to shoot a honey buzzard. This is a ritual which is supposed to ensure that their wives will be faithful; ironically, it entails leaving them completely to their own devices for the duration of the hunting trip.

But jealousy is certainly not only the prerogative of the male. Women, too, feel jealous, but for different reasons. They are more concerned about their mate diverting care, attention and material resources to another woman. So while men are more worried about their partner having casual sex with another man, women are more concerned about an emotional involvement which may lead to the disastrous scenario of abandonment and poverty. I understand that Greek women whose men have slept around are not pitied or degraded; enduring the infidelity is entirely acceptable, and may even be expected, as it's seen as being far better than the alternative of having to fend for the children on your own.

Nowadays, many single mothers do not need to be supported financially, or even emotionally, by a man, but the modern world is not where our instincts were forged. Would these instincts have made sense on the savannah? Can we imagine jealousy as a driving force of natural selection, an emotion that gave bands of pre-humans the edge over their competitors? Yes, although it's fair to say that our ancient instincts do not hold sway over each and every one of us. There are couples who indulge in 'wife-swapping' where the men actively encourage their mates to be casually promiscuous. There are women who divorce their husbands after discovering one-night stands, even if there was absolutely no emotional involvement.

These are all exceptions, but even they are touched by these age-old passions. Our instincts are whispering voices in our unconscious that push and pull our behaviour a little bit this way then a little bit that way. Studies across many different cultures show that sexual jealousy is universal. David Buss has shown that men and women *do* react differently to extra-marital sex and to emotional involvement; the same results have been recorded for people in various countries and cultures, from Holland to Korea to the US.

But for many years, sexual jealousy was thought not to be a universal human instinct, rather a product of our own Western sexual repression, hang-ups and neuroses.

The truth revealed?

In 1925, Margaret Mead, a legendary American anthropologist, set out to chronicle the lives of hunter-gatherers on the island of Tau in Samoa. Her account gave a picture of lives that were starkly different from our own in the affluent West. Mead described an extraordinarily peaceable culture in which violence was extremely rare, equality of roles between the sexes was the norm and sexual relations were fluid and easygoing. Jealousy, and the rage and violence that are occasionally the end result of it, was, she claimed, non-existent.

As a youngish teenager, I remember picking Margaret Mead off my parents' bookshelves. She was highly fashionable in middle-class thinking and I imagine my family greatly approved of the perfect island idyll she seemed to describe. Here was a society in which human beings could live without the destructive tendencies of our own struggles and sexual tensions,* and her account matched the moral ambitions of those theorists who believed mankind was, in some way, 'perfectible' – in other words, that society and culture could be moulded in such a way that our potential for a peaceful, creative and moral existence would be achieved. They had been waiting to hear about a society like that of the Samoans. Murder was practically unknown. There was a communal sharing of responsibility and authority. Feminists were especially pleased to hear of a culture that did not rely on entrenched sexual division of labour. In fact, the Samoan hunter-gatherers sometimes seemed completely to reverse traditional Western gender roles, and men were often the designated baby-raisers and homemakers.

Mead's views were embraced by people eager to believe that sexual division and violence were cultural phenomena, something for which we are not necessarily genetically hard-wired. Mead seemed to be saying that we do not need to be the way we are. It is only our Western way of life that blemishes the 'tabula rasa', the

* As the reader may imagine, as an adolescent I was pretty impressed too. I couldn't think of a more attractive place to visit than one where I might indulge in the sexual delights that apparently were so freely on offer.

blank slate with which we are all born, leading to violence, moral disintegration and sexual confusion.

If Mead were right, then one would have to be very careful about proposing any kind of universal sexual instinct. Even if there were just one or two cultures which showed significant differences from our own, and if that culture was shown to be consistently devoid of the vices that afflict the rest of humanity, then we would have to rethink. Our behaviour could still be cultural rather than instinctual. Sexual jealousy is a common trait in many human cultures, but could it still be an emotion we 'learn', something which is not part of our genetic programme but part of a cultural and social tradition of sex and family life?

Many years later, in 1983, the Samoan bubble burst. An Australian anthropologist, Derek Freeman, returned from an intensive study of the Samoans with very different conclusions and the belief that Mead's earlier work amounted to little more than fiction. Violence there, he said, rather than being almost completely unknown, was worse than that of some inner cities in the United States, with higher rates of murder and rape. There was no sign of a free and easy sexual culture, and sexual jealousy was rife. Mead's island idyll was just as violent and divisive as any Western city. Freeman claimed that Mead had been purposefully 'anti-evolutionary' in the way she had ignored biological factors in favour of her theory of cultural determination of sexual roles. New evidence emerged that Mead had been a little loose with her descriptions, for instance characterizing one particular group of men as gentle and unaggressive while one of her co-workers had recorded that many old men 'claimed one or more war homicides to his credit'.

Freeman took much criticism for his re-evaluation of Mead; the anthropologists were not quite ready to overturn her idealism and utopian view of human nature. But the orthodoxy was eventually overcome, and now there is a grudging acceptance that some things are the same the world over. Sexual jealousy is, in my view, fairly close to the top of the list.

As for Mead, it is unclear why she missed some central features of the sex lives of the Samoans. Was she duped by her 'informants', those insiders who were recruited to spill the beans on their lives and on others'? Other researchers have claimed that the Samoans have a fondness for joking when talking about sexual behaviour; possibly Mead mistakenly took their quips for the truth. Maybe she found what she wanted to find. Some may say she was a dreamer, but she was not the only one.*

Violent love

It is an obvious but important observation that jealousy begets violence. I vividly recall an early experience as a houseman at The London Hospital in Whitechapel. In those days, the days of the Kray twins and other violent gangs, much of the East End, especially around Brick Lane and Cable Street, was pretty dangerous. I was working one evening in what used to be called the Receiving Room (RR) at the time, now called Accident and Emergency.

About 2 a.m. a heavily bleeding male of about forty-five years was found slumped between two chairs in the waiting area. It wasn't obvious how he had got there, dressed as he was in a blood-stained shirt with no trousers and dirty bedsocks. Once placed on an RR couch it was immediately evident that he had a nasty wound in his right calf and the back of the thigh of his right leg. Seeing as he refused to speak at all, it was not too clear how he had got the wound. But my registrar was experienced in pheasant shooting – goodness knows with what kind of sportsmen he usually associated – and he recognized a shotgun injury.

At around 2.45 a.m., a hysterical woman came into the RR enquiring after her husband, Mr Smith. She said she thought he was here and that he had hurt his leg on the Aldgate night-bus. She

* Mead remained nonetheless a firm advocate of free and open debate. When, in the mid-1970s, Edward Wilson published his book *Sociobiology: The New Synthesis*, many social scientists were horrified at his suggestions that biology could explain human behaviour. At a gathering in Washington DC in 1976 they tried to get his books and others banned in schools and universities. Mead, still America's most famous anthropologist, took the stage and made an impassioned defence of the right of the sociobiologists to publish their views, likening her colleagues' actions to medieval book-burning.

claimed she was our patient's wife once she heard his voice through the curtains of his cubicle. We reassured her that his condition was stable and that the wound needed dressing and orthopaedic care in the morning. She immediately left for her house, giving a Cable Street address. Twenty minutes later another Mrs Smith came in, enquiring after her husband. She seemed as hysterical as the first woman; we were increasingly confused when she claimed the same injured man as her husband. After a few minutes she left suddenly. Thirty-five minutes later, another man was brought into the RR. We got a fleeting glimpse of the woman who had accompanied him to hospital before she fled, but she bore a strong resemblance to the first Mrs Smith we had seen earlier that night. This man, who also refused to give his name, did admit to living in Cable Street. What was all too apparent were the extensive gunshot wounds to the buttocks and the back of both thighs.

The rather bizarre puzzle was finally solved at 5.15 that morning when both women who had called themselves Mrs Smith were admitted within ten minutes of each other suffering various abrasions, a black eye apiece, in one case a rather nasty split lip and in the other a broken collar bone. As soon as they saw each other in the waiting room they started throwing the chairs and were only prevented from doing each other fatal injury when the first Mr Smith limped into the waiting area and separated them.

Cable Street is a narrow street at some points, and the two men lived opposite each other. At some time after midnight, Mr 'Smith' (A) was in the front bedroom of his neighbour's house. He was preparing for a comfortable night away from home on the understanding that Mr 'Smith' (B) was doing a job in Liverpool. Unfortunately, when Mr 'Smith' (A) drew the curtains in anticipation of the delights provided on the odd-numbered side of the road, he could see Mr 'Smith' (B) on a different kind of job in his own bedroom. To add insult to intended injury, he could see he was wearing his birthday present – white bedsocks. Knowing there was a shotgun in the wardrobe, he loaded it and fired it through the window, hitting his victim in the calf of the leg closer to the window. He then returned to the work he had originally in mind.

Nemesis came later that night when his wife returned the compliment, proving a rather better shot under what were surely unusually trying circumstances.

Anecdotal evidence and everyday experience suggest that domestic violence is a common consequence of sexual jealousy. The statistics back this up; 13 per cent of all homicides involve the husband killing the wife, or vice versa, and male sexual jealousy is the overwhelming cause of these, even in the cases in which the wife kills the husband. Even Margaret Mead's Samoans eventually turned out to be typical in this respect. Studies in Baltimore showed that over a single year 81 per cent of husband/wife murders were motivated by jealousy on the part of the husband. In the Belgian Congo, the figure averages around 93 per cent. In Canada, 85 per cent of domestic spousal murders in one year were 'caused' by the 'sexual behaviour' of the wife. Note that jealousy need not be based on actual infidelity, just the *suspicion* of infidelity.

And of course, while those who actually commit murder are doubtless at the extreme fringes of the spectrum, the bounds of sexual jealousy extend into everyday life for almost everyone at some point. Surveys carried out by David Buss revealed that nearly all men and women had undergone at least one highly intense experience of jealousy relating to a partner; 31 per cent admitted that their sexual jealousy had sometimes been difficult to control, and of those, a further 38 per cent revealed that at times their jealousy had led them to want to hurt someone.

Most people appear to consider that unfounded jealous suspicions and consequent rages in others are more 'out of control', verging on the pathological. It seems that fury resulting from the actual discovery that infidelity has taken place is often intuitively understood. The law in Texas as late as 1974 recognized the killing of the lover of one's husband or wife as 'justifiable homicide', so long as it happened at the scene of the discovery. In fact, this particular law, like many others around the world, went so far as to propose that it was quite 'reasonable' for a man to act in this way, having discovered his wife in the arms of another man. It would seem that cheating hearts are the ultimate when it comes to extenuating circumstances.

Though sexual jealousy can get completely out of hand and lead to murder, for most of us this ancient feeling does serve an important purpose. On an everyday basis in our relationships, jealousy probably really does help us to cope with a whole host of small but real threats. Unconsciously or otherwise, it's probably what motivates us to ward off rivals with nasty stares or even threatening words, or to shower our partner with affection when we feel he or she may be on the verge of straying. Jealousy is a double-edged sword: it can keep a long marriage stable over many years, or it can drive a man to beat his wife. Maybe this is how men came to be naturally more aggressive?

The big T

Although testosterone is traditionally thought of as the male sex hormone, you may be surprised to hear that women also have it pumping through their veins, though in nowhere near the same amounts. In fact, women produce a trifling amount of this hormone every day, just enough to ensure normal ovulation; adult men produce some twenty times more – around 7mg per day. Those high levels of testosterone are one reason why men are generally more aggressive than women. If we look at our primate cousins, the chimpanzees, we find that males are also more aggressive than females. Developmental psychologists have long noted that even among young baby boys there is a definite trend to more aggressive behaviour and more 'rough and tumble' play as they grow into toddlers.

It seems probable that as our ancient ancestors evolved over millions of years, the males – who were sometime hunters, sometime scavengers and always the defenders of the group – evolved some successful biological mechanisms to fulfil these roles. We know that with violent predators to fight off, hominid males had to be strong and powerful. We also know that testosterone builds muscle bulk and thus physical strength. But aside from its effects on muscles, a man's daily production of testosterone also has a huge impact on many aspects of his behaviour, not least his sex drive. Testosterone certainly accounts for the higher sex drives of men and may perhaps have evolved hand in hand with their desire to impregnate as many women as possible in order to spread their genes. Testosterone gave them the power to go and do it, so to speak.

Studies have demonstrated that a male monkey, when shown images of another monkey having sex, will exhibit a rise in testosterone levels by perhaps 400 per cent. Among adult rhesus monkeys who are socially dominant in their troop, there are much higher levels of free testosterone which subsequently fluctuate in line with the rise or fall of their social rank. Research on testosterone levels in human males attending football matches shows that the level of the hormone rises dramatically in those men whose team have just won, as opposed to the crestfallen supporters of the losing side whose mean levels of this hormone are likely to fall.

These findings are too consistent to be due to chance; Dr Allan Mazur and his colleagues working at the University of Nebraska, and more recently at Syracuse University, have found strong correlations between levels of male hormone and dominance. They measured both cortisol (the stress hormone) and testosterone in six university tennis players before their matches. In general, levels of testosterone rose before the matches and those players with the most positive feelings about their match had the greatest rise. After the match, winners had higher levels of the hormone than the losers; it is particularly interesting that winners then tended to have higher than average levels before their next match, too, while losers had lower levels. Cortisol levels, on the other hand, were hardly affected, but subjects that were seeded highest tended to have the lowest levels. Perhaps it is less stressful knowing you are at the top?

Being a competition chess player myself, and knowing how aggressive most match players tend to be, the study of Mazur's I like best concerns this ancient, competitive game which has always appealed so strongly to male instincts. He and his colleagues measured testosterone levels in closely matched men. After the game, the winner experienced massive rises in blood levels of testosterone. In the guys who lost, testosterone plummeted. So even in chess, status is linked to male hormone changes.

The darker side of men and sex

The male sex drive particularly intrigued Leonardo da Vinci. 'The penis', he observed, 'does not obey the order of its master, who tries to erect or shrink it at will . . . instead the penis erects freely while its master is asleep. The penis must be said to have its own mind.' This view of men implies they are literally out of control; they are sexual automata, whose only ambition is to impregnate as many women as possible.

A man's libido has a darker side, especially when sex is accompanied by physical coercion. The traditional sociological view of rape is that it is a pathological form of behaviour, a crime committed by dysfunctional individuals, a maladaptive crease that should be ironed out through moral and cultural education. The emphasis, especially from the feminist perspective, is on violence rather than sex. The force behind rape, they say, is an urge to harm someone and be in control of them rather than an urge to procreate.

Most people can surely never have imagined that rape could be described as 'useful', from the point of view of human evolution, but controversially that is what some researchers have recently suggested. While making it very clear that their theory does not provide any moral justification for rape, they say that historically it is actually possible that it could have been in a man's interest to employ force as a means of coercing a woman to have sex, alongside the normal methods of consensual sex with short- or long-term partners. Rape, they claim, could be a fruitful strategy from the point of view of a man's genetic inheritance.

Rape? A useful male strategy? There were, understandably, some fierce reactions. No normal man wants to know he may instinctively harbour a desire to rape women. No woman wants to know that rape is in any sense a 'normal' form of human behaviour. And the claims appear to have serious flaws. For one thing, many rape victims are too young or too old to bear children, and sex without the promise of conception is of no use to evolution whatsoever. More importantly, after a rape there is little chance of mother and father taking care of the infant together, and therefore the chances of survival of the resulting child must be significantly reduced.

The chance of pregnancy after rape is a controversial issue. There are some studies which bizarrely indicate that, for some unknown reason, the chances of a woman conceiving from a single act of rape are more than twice those of a woman who engages in a single act of consensual sex. Some scientists have suggested that rape increases secretion of stress hormones in the body and that these may trigger ovulation if the rape takes place somewhere near the middle of the menstrual cycle. Is this some throwback to the time of the caveman? Certainly this statistic, if indeed it is true, provides a possible evolutionary reason for rape, and suggests that hardcore feminists might not have been so wrong when they said that every man is a (potential) rapist, rather than rapists simply being dysfunctional or pathological individuals. But at present the statistics are too blurred to make firm assertions. As other authorities state that from their observations pregnancy is less likely after intercourse which is not consensual, it would be unwise to draw definite conclusions.

Laws and cultural and moral norms are put in place to repress and contain our 'animal' instincts. What happens when that legal and moral code breaks down? Does instinct take over? It is true that amid the chaos and lawlessness of war-torn Yugoslavia, rape became endemic, particularly the rape of women of the 'enemy' by occupying soldiers. Rape is a feature of many such conflicts, a 'tradition' that goes hand in hand with the plunder of money, gold and other valuables. Women, too, are treated as objects to be looted. Under conditions of war there is far less chance of punishment or reprisal, so perhaps an absence of society means an increase in sexual violence.

However, where there is a strong societal pressure, among traditional village cultures for instance, rape within the group is met by fierce revenge or severe punishment – as it is in practically all human civilizations. But among hunter-gatherer or tribal cultures who wage war on neighbouring groups, when rape is committed *outside* the group it is often called something different, and this kind of rape is not seen as a crime. Like the occupying forces of Western nations, there is at some level an acceptance of rape.

Among the warring bands of early humans, it is just possible that rape was more commonplace than we think. It is unpleasant to consider it, but rape could even have become part of the male sexual instinct. This, of course, in no way condones this aspect of human behaviour. The tentative application of an instinctual element to a particular behaviour should in no way change our attitude to its consequences for other people. After all, we all have a choice in how we behave. But in evolutionary terms the question of mother and father taking care of the baby after rape would have been irrelevant as the baby would have been raised in another group entirely. The genes that predispose a man to commit rape in these situations would become more and more successful. If this is the case, perhaps rape is not just a product of a few diseased or dysfunctional minds but, as difficult as we find it to admit, an actual part of our human instinct. All the more reason to be on our guard; rape may be instinctual, but that does not make it understandable or acceptable. In later chapters we shall examine other aspects of human behaviour that are equally disturbing and which also have their roots in our evolutionary past.

Research by my own team gives a contrary but slightly relevant view. We wondered whether sex which the female partner found pleasurable improved the chances of her having a successful pregnancy. I have long wondered whether the female orgasm plays any significant part in conception, so my team set up the following study. With fully informed consent and giving cast-iron assurances about their privacy, we asked several hundred infertile women about their regular sexual experiences. The questionnaire was designed by psychologists and statisticians in such a way as to be likely to give statistically valuable information. Two hundred women with a known cause for infertility (about half of them had blocked Fallopian tubes) were compared with a matched group (matched for age, health history, body weight etc.) of two hundred women who had no known cause for being infertile. In all cases their partners had sperm counts within the normal fertile range.

What we found was that women who were infertile with no apparent physical cause less regularly reached orgasm or reported experiencing less pleasurable sex on frequent occasions. The control group, with a clear cause for infertility, mostly reported more sexual satisfaction. What, incidentally, was striking was that women with tubal blockage, all of whom had had tubal infection as the cause, reported the highest incidence of orgasm. One possible explanation is that female orgasm assists the transport of sperm through the uterus and into the Fallopian tubes. And of course, with the seminal fluid may well come some micro-organisms, normally harboured in the vagina where they don't give rise to damage; in the Fallopian tubes they may well cause scarring. More work needs to be done, but this study raises several interesting questions. If sex which is pleasurable to the woman does improve the chances of conception, it must be in the man's interest to ensure that his partner reaches orgasm whenever possible. This may be why men seem to enjoy sex most when their women have intense sexual experiences and why men generally take the greatest pleasure in female orgasm. Could this, too, be an instinctive phenomenon selected from our time as early hominids?

A monopoly on women

Whether or not we choose to believe that rape is, at some level, part of the male human instinct, we do know for sure that men 'lay claim' to certain women as their own. There is a strong proprietary nature about the male sexual instinct, and this ties in with that strong need to avoid being cuckolded. Some people believe that because women must invest so much more of their biological resources and energy in bearing and raising children, they are, in effect, a limited resource themselves. Men have evolved to monopolize females, just as any animal would contest its territory in the struggle for survival, for instance.

But with man's evolutionary need to 'lay claim' to a woman to ensure her availability as a child-bearer and her fidelity, history and culture have combined to lend a hand over the ages, translating that monopoly into what has been considered to be a man's *right*. Women have often been thought of as representing commodities, and have been treated as such. Binding the feet of women was popular in China from the Middle Ages, a method that literally prevented them from walking away, and in the meantime caused permanent and painful deformity. In medieval times, it has been claimed that women were frequently locked or even welded into a painful and restricting chastity belt to prevent any chance of their infidelity; when their knight returned from the latest Crusade, the belt would be unlocked and the man's 'protected property' inside released.*

Other methods are less overt and rely on a social or economic system that reinforces the idea that women 'belong' to their husbands. In their essay 'The Man Who Mistook His Wife for a Chattel', Margo Wilson and Martin Daly give the example of the English ritual of a man selling his wife at market, which apparently died out only around 1900. The husband would pay a fee to the market owner and then lead his wife up onto the stage where she would be sold to the highest bidder. The auction had a more symbolic role of officially marking the divorce of the couple, and she would typically be sold to the man who was already her lover. Yet, as Wilson and Daly point out, the ritual does symbolize the notion of a wife as a piece of property.

Civilizations from ancient Babylon to present Western society codify and regulate the marriage bond in terms of male ownership of the female. Even the modern wedding ceremony involves the father of the bride 'giving' his daughter away. Adultery and marital law has generally revolved around the idea that the husband has rights to his wife as he does to any other piece of property. If someone else takes her away, he deserves compensation, as he would if he were the victim of a theft.

And often the male lover pays the highest price. In ancient India, if a man had sexual relations with the wife of his guru, he was forced to sit on a burning-hot iron plate and then made to chop off his own penis. When Loretta Bobbit did the same to her husband's

* These reports are almost certainly exaggerated myth. The first plausible chastity belt appears in the *Epic* by Marie de France (c.1180). It narrates how the knight Guigemar bids goodbye to his lady love amid her tears that if he were killed, she would no longer want to live. She ties a knot to the end of his shirt, one that could only be opened by force, and begs him to be faithful. In turn, Guigemar takes a girdle knotted in a peculiar fashion and ties it around the naked body of his lady. Guillaume de Machaut, the fourteenth-century musician whose songs are enjoying a considerable revival now, gives a similar example of a symbolic chastity belt, and in the fifteenth century it seems the women of Florence in Italy voluntarily wore a piece of metal to discourage possible attacks on their virtue, rather than to keep them from illicit liaisons while their warlords were away.

appendage, she was unknowingly reviving an ancient and sacred tradition.

Men as resource providers

These practices, some of which clearly verge on sexual slavery, were the products of human civilization over the past ten thousand years; but on the scale of evolutionary change, this period of human history is merely the blink of an eye. These practices were not factors in determining human instinct. To understand instinct, we need, as always, to go back to the time on the savannah.

We have assumed that men tended to provide for the mother and child throughout the pregnancy and beyond, when the newborn was at his or her most vulnerable. Without men to provide food and protection, surely women would have had their chances of successfully rearing a healthy infant significantly reduced? Remember that over the past thirty years or so the image of Man the Hunter has been usurped by Man the Scavenger and Gatherer. In fact, women might have done much of the gathering, just as they do in some modern hunter-gatherer cultures like the !Kung San of the Kalahari Desert. San women consistently gather plentiful supplies of mongongo nuts (rich in protein) while their men are trying, and often failing, to hunt big game. The men in modern hunter-gatherer cultures in New Guinea also often spend a large amount of time in a fruitless search for kangaroos while their wives and children are bagging fish, insects and other goodies back home. Perhaps women were relied upon to bring in the staple foods that kept the group alive; even when women were pregnant and suckling their infants, they were still capable of gathering food. Perhaps the truth is that the men did not bring home the bacon after all.

Our ancestors' life on the savannah presents many unanswered questions. No-one can be certain that men were indispensable when it came to hunting, gathering or looking after children, but we do know that compared with most other species our newborn babies are extremely vulnerable and need a vast amount of care and attention. The mother almost certainly had to spend a great deal of time nursing the baby – breast was not only best, breast was the only option. We can be pretty sure that a band of hominids in which the menfolk played a useful role in providing food, water and other resources would have had an advantage over their competitors. We must also consider the size and strength of males compared with females; males were better suited to warding off attack from predators as well as fighting off intruders from another band of hominids. It seems harsh to tag all early human males as feckless layabouts with no role to play in the success of the family unit.

Sexual stereotypes have a long and distinguished history. What of the heat-seeking missile that is the male libido? The aggressive, competitive male and the coy and reticent female are entrenched stereotypes, and their familiarity makes them difficult targets for criticism. But just because most modern cultures sustain those stereotypes, it doesn't mean those behaviours are necessarily genetic, or an adaptation from the savannah. Our sexual habits could simply be products of human culture; a tradition once adopted is hard to shake off. Culture becomes entrenched, like culinary traditions, or religion; customs and conventions like brewing beer or worshipping Jesus become part of life for generations. But no-one would ever imagine they are adaptive forms of behaviour, forged through natural selection.

Monkeying about

Some years ago most species of monkeys and apes were thought to share similar kinds of sexual behaviour to humans. The males were described as acting like drunken yobs in a singles bar, competing for sex, showing off their brute strength and their virility and regularly fighting over women. The females, on the other hand, were thought to be shy and sexually circumspect. In the 1970s, Meredith Small and other mainly female primatologists cast a sceptical eye over this version of events. They found that the social and sex lives of primates don't always fit with the traditional view of male and female behaviour. Among baboons, for example, closely knit groups of females were found to have central roles in the social set-up of the troop, including control over which males were granted sexual access to the females. In other species, the females, during their fertile period, were found to initiate sex, sometimes by running up to a male and waving their reddened behinds in front of him – which seems a fairly unequivocal invitation.

There may be a good evolutionary reason for this. If neither male or female is quite sure who the father is of any given offspring, then the mother could get attention, resources and protection from a number of males, all of whom are anxious to protect their investment. (Equally, though, they may get no support at all from any of the males, all of whom are anxious *not* to raise someone else's baby.) Meredith Small thought it might also simply be the fact that sex is pleasurable for the female so she desires it – a simple explanation that may be one of the overriding factors when it comes to sex. Whatever the reason, previous researchers – mostly male, it should be said – had completely failed to look for this kind of female behaviour.

More striking was a report in *Nature* in 1997 about the sex lives of female chimpanzees. According to DNA analysis of a group of fifty-two chimpanzees in the Tai Forest in the Ivory Coast, more than half the chimp babies had been fathered by males outside the group. The researchers proposed that females, who had been thought to mate mainly with the alpha males within their own community, had been sneaking off at moments of peak fertility to have sex with males from rival chimpanzee factions. From the point of view of producing a healthy baby with a good mix of genes – an infusion of 'new blood', as it were – this is a reasonable tactic. But the research upset some long-standing assumptions about chimpanzee sex lives. If females strayed outside the group then alpha males were effectively cuckolded. Their hard work in climbing to the top of the male chimpanzee tree was being undermined by outsiders. The chimp females were the males' worst nightmare – promiscuous, disloyal and deceptive.

The results were widely disseminated and the conclusion was quickly accepted, but four years on a new study was published, this time of a group of chimps in Tanzania. The researchers found that males within the group had indeed, as would have been previously expected, sired all fourteen infants. Then the authors of the Ivory

Coast study admitted that their DNA analysis was possibly faulty in part as a result of the difficulty of obtaining DNA samples from chimps in the wild. The researchers were often forced to analyse hairs shed by the chimps, which contained very little usable DNA material. The later study was based on DNA from faeces, a more reliable source. It also employed automatic DNA sequencers that performed the comparisons much more accurately.

Primatologists around the world breathed a huge sigh of relief. Their years of painstaking observation of chimpanzees had turned out to be correct after all, and the alpha males regained their place as high-scoring Lotharios who took the lion's share of mating opportunities: half of the chimp offspring were found to be fathered by high-ranking males. Still, the study showed that the females retained the power to choose their mates. Half of the infants in the Tanzanian group were sired by low-status males, and the females were instrumental in allowing them sexual access. The power wielded by the chimp females may have a parallel in human sexual conduct. Despite the relative power and position of men in our society and over the ages, controlling access to sex is a powerful position to be in, and women are more often than not the gatekeepers.

The end of feminism?

As a member of the Athenaeum Club* in Pall Mall, I often climb the stairs to the library with the remarkable portrait of Charles Darwin on the wall. The Athenaeum was founded in 1824 as 'an association of persons of literary, scientific, and artistic attainments, patrons of learning' – or, as more usually put, 'men of the mind'. 'Men' notice, not women. Darwin, perhaps unsurprisingly for a man of his class and era, had no truck with sexual equality. He believed that a paternalistic society was as natural as the movement of the earth around the sun – 'the masculine force always predominates'. He set the tone for the misuse of evolutionary theory as a means of

* I am very pleased to say that finally the Athenaeum has voted overwhelmingly to admit women, but it took over 170 years.

justifying male oppression and control of women (if Darwin were alive today, I sincerely hope he would draw different conclusions), so it is easy to see why many feminist thinkers have greeted the recent wave of evolutionary psychology – or 'evolutionary sexism' as they would see it – with a hail of rhetorical bullets.

Some of them condemn out of hand the project of examining the genetic basis of sexual behaviour. They believe that even to pursue this line of inquiry is a gross affront to women, and that it gives support to the worst kind of sexual conservatives, mainly men, who believe women should be no more than maidservants and breeding machines. Other writers and academics are more thoughtful, presenting arguments that make the evolutionists sit up and think. Paul Ehrlich, for example, a well-respected evolutionary biologist and psychologist, has attacked the notion that women are hard-wired to be sexually 'coy' and that men are destined to be more promiscuous. 'Women, like men, evolved to be smart,' he says. 'They certainly don't need to be rocket scientists to understand that they make a bigger potential commitment to each sex act than do men.' That alone, rather than an evolved tendency in men to reproduce more, could explain differences in attitudes towards fidelity, and thus differences in the behaviour of men and women.

According to Ehrlich, our ability to reason overcomes instinct; the savannah is overtaken by conscious deliberation; rationality takes the place of evolution. But for those who believe in the evolutionary roots of sexual behaviour, there is no acceptance that reason has anything to do with what appears to be an ancient and universal pattern of sex differences. The argument can get extremely heated. Camille Paglia, typically, takes no prisoners. Maybe she should have the last word. 'If middle-class feminists', she says, 'think they conduct their love lives perfectly rationally, without any instinctual influences from biology, they are imbeciles.'

chapter**four**

A Family Affair

The mating instinct

There is a fish that could be considered the patron saint of radical feminism. The female pipefish mates with a partner who fertilizes her eggs in the normal way, but then the male is literally left holding the babies. In this topsy-turvy species, it's the males who have a pouch in which to carry the brood, and the females, as if trying to make up for the countless males across the animal world who leave their partners in the lurch, who carelessly wander off in an attempt to seduce another male with an empty pouch. And no, she didn't phone the next day, either.

It's a perfect twist in the sexual story. Females scour the territory, hungry for sex, while males selectively pick and choose the best genetic bet for their next pouch-load of eggs. Female pipefish get to play John Malkovich's Valmont to the male's Michelle Pfeiffer; the seducer becomes the seductress, without caring about the consequences.

In many species of pipefish the females are more brightly coloured than the males, advertising their fertility and health. The males are choosy about who they allow to put eggs in their pouch. The male pipefish is the brake on sexual activity; the female just wants to 'impregnate' as many males as she possibly can, and she has to compete for the privilege.

It all depends on who makes the most investment in the offspring. In most species, the females get saddled with the work and the males have to compete for females. And when there is a high level of competition among the males there are some interesting consequences, particularly with respect to size.

Consider a monkey, the mandrill. The male makes a big show of his masculinity; he has a brightly coloured face and he's three times the size – and weight – of a female. Male hamadryas baboons and gorillas are also outsized compared to the female. But the species for which size really does matter is the elephant seal. The male is a whopping seven times the weight of the female (not that the females could be called petite themselves). From the female's point of view, sex with a male elephant seal must be like having a lustful liaison with a Sumo wrestler – and nearly as dangerous.

This mismatch in male and female body sizes and forms is actually surprisingly revealing about the sexual habits of any particular species. Sexual dimorphism, as it's officially known, is at its strongest among species that mate according to the harem system: one male and a number of females.

Each dominant male mountain gorilla is typically surrounded by a generous handful of females in any one group. There is bound to be fierce physical competition for the right to mate with a large clutch of females. In the past, those gorillas who just so happened to be bigger and stronger ended up having an edge in these macho contests, so the evolutionary pressure for the males to increase in size was born. Sheer bulk is clearly important where physical violence is concerned, but it's certainly not the only weapon in the armoury. It turns out that some animals' characteristics which we may assume have evolved as protection against predators are in fact more commonly used to fight off other males in the quest for sex. Deers' antlers are a good example of this.

Among elephant seals, most calves are fathered by less than 5 per cent of the males. They fight, just like Sumo wrestlers, by rearing up and trying to knock each other over. The male who wins the fight gets the girl and the ultimate prize of siring offspring and passing on his genes, while the skinnier contenders have to wait, offspring-less, in the wings. Survival of the fattest, in this case.

So what about humans, who have so many of their genes in common with all other animal species? Human males also tend to be slightly bigger than females. That simple fact actually reveals something fascinating about our hominid ancestors. Our ancient male forefathers probably also had to fight for mating rights, suggesting that early human groups were commonly practising 'polygyny' – in other words, a proportion of the males would monopolize all the females. If everything else is equal, and if the theory is correct, we were living in a mild human version of the harem system, although not nearly to the same extent as the elephant seals or gorillas. But it is likely that there was some physical competition between males for access to the females. Men fought over women, inevitably, as they still do.

For the Yanomamo people of the Amazon jungle, fighting and warfare is a way of life. Up to one quarter of all Yanomamo men die in tribal battles, but the surviving heroes often then go on to be incredibly prolific in the mating game. The original founder of one group of villages claimed eight wives and forty-eight children as his own. As luck would have it, his sons also turned out to be good at strutting their stuff; in the end, some three-quarters of the entire population were descendants of the original founding male. The anthropologist Napoleon Chagnon spent most of his life observing and studying the Yanomamo people. In an attempt to understand the men's strong leanings towards violence and fighting, and why they have among the highest rates of homicide in the world, he asked a group of them why it was so important in Yanomamo society. There were astonished looks all round. 'What? Don't ask such a stupid question! It is women! Women! We fight over women!'*

* But adverts like this suggest that women in the West may now be taking the whip hand: 'I had a wonderful time at the submachine-gun course. The instructors were talented, knowledgeable, courteous and safety-conscious. This course is a must – especially for women!' *Michelle Martin, kindergarten teacher, on the Nevada Pistol Academy.*

An ethnographic survey of 849 human societies across the globe, including all traditional tribal and hunter-gatherer groups, showed that 708 of these are polygynous or allow the powerful or wealthy males to indulge in polygyny. It seems that, from a strictly evolutionary point of view, the competitive nature which pushes a man to pursue power and status is strongly related to sex – and the number of women he can have sex with. That certainly puts modern politics into a whole new light, at any rate.

The anthropologist Laura Betzig maintains that tyrants and despots have kept harems whenever possible. When hunter-gatherers settled down around ten thousand years ago and started to build villages, cities and kingdoms, there was greatly increased centralization of power. Leaders tended to rule by excessive authority and sometimes terror, so there was much more potential for polygyny. The record stands with an Indian emperor called Udayama, who ruled in the fifth century BC and supposedly kept a harem of sixteen thousand women. (Even assuming Udayama had sex with two women every night, it would still have taken him almost twenty-two years to sleep his way through the entire harem.) One of the earliest biblical examples is Ahasuerus, King of Persia, who certainly flaunted his power by having as many attractive women in his harem as could be found throughout the 127 provinces of his empire. But he was still susceptible to female wiles, and his favoured (excessively beautiful and very young) queen, Esther, was able to exploit this vulnerability.

The Romans, despite having a reputation as monogamous and faithful lovers, kept slaves for the purposes of sex. Child slaves who were raised in the master's home were called *vernae*, and traditionally historians have assumed their fathers were also slaves, but Betzig argues these children were almost certainly fathered by the master of the house. Children born of these illicit liaisons were often treated well, educated alongside the master's 'legitimate' children, and even sometimes inherited parts of his estate.

There was a tradition in medieval Europe concerning a right called *jus primae noctis*, or 'right of the first night', an ancient privilege which gave the lord of the manor the right to bed a peasant's bride on their wedding night. (It is unclear how often this privilege was claimed; possibly not all servants were as lucky and as wily as Susanna in her dealings with Count Almaviva in Mozart's *Marriage of Figaro*.) And while the lord would generally have a wife, his household was set up as an unofficial harem of servant girls and lower-class young women for whom the manor served as a haven from poverty. Count Baudoin, a cad of the thirteenth century, was typical, according to Betzig. He fathered twenty-three illegitimate children under his own roof and saw to it that his bedchamber had interconnecting doors to the servant girls' quarters, to the rooms of adolescent girls upstairs and even to what was known as 'the warming room' – the place where the infants of the house were suckled.

As the structure of Western society changed, as the landed gentry began to lose power and wealth and as life became more democratic and egalitarian, polygyny on this scale died out. Household staff were still employed by aristocratic families and the landed gentry, of course, but their numbers decreased dramatically, and they got on with the more mundane tasks of polishing the silver and serving the tea. Still, those who have seen the film *Gosford Park* will recognize that medieval male aristocratic practices and relationships below stairs did not die out.

But societies, ancient or modern, don't always order their affairs according to human instinct. They might have invented ways, means and even traditions to try to repress instinct. The institution of marriage may in fact be a reaction to our innately unfaithful tendencies that are, for reasons we'll consider later, thought to be undesirable or destructive.

Why size matters

The physical clues to human sexuality give us a mixed message. Size does matter, and I don't only mean the relative height and weight of males and females. Believe it or not, the size of an animal's testicles can tell us a lot about its sexual practices. Chimpanzees have extremely large testicles and produce prodigious amounts of sperm. The reason for this is that chimps have a very relaxed attitude to sex. Mating is relatively unregulated and there seems to be no awareness of the paternity of any of their offspring. Chimps, seemingly, could be regarded as the originators of the free-love hippy commune. When a chimp female is in her fertile 'oestrus' period, she has sex with a large number of males, not just the alpha male, who may try but often fails to dominate the mating game. Therefore each male's sperm has to compete to fertilize the ovum, and the more spermatozoa a male produces, the more chance he has of becoming a father – literally by flooding the female's genital tract with sperm.

Silverback gorilla males are large, aggressive and scary beasts, but they have very small testicles. This reflects the fact that Silverback alpha males take possession of harems of females and are secure in the knowledge that females rarely sneak off for an illicit rendezvous in the woods. Gorillas rarely have sex, because for those males with a harem sexual access is guaranteed; therefore they seem to need only a small amount of sperm.

Measured as a proportion of body weight, human testicles are four times the size of a gorilla's but less than a third the size of a chimpanzee's. What may we infer from this?

Human males tread the middle ground. They've acquired testicles big enough to combat a moderate amount of sperm competition from other males, but they certainly don't need the heavy artillery of the chimpanzee gonads. Taking these measurements as indicators of our ancient sexual practices, we were probably a mixed bag. We had an element of the harem system, judging by the relative size of the male and female body, but we were also equipped for occasional female promiscuity too. It also seems that testicular size varies slightly across the human species. Even allowing for variations in body size, one study has shown that Japanese and Korean men tend to have rather smaller testes than Europeans, and the Chinese weigh in at half that of their Danish counterparts.

Penile size is also probably very important in primates. Once hominids were standing upright on two legs, their genitals would have been much more apparent to other members of the group. Moreover, the human penis is big. The average size is around five to six inches, compared with three for chimpanzees and perhaps half that in gorillas. A number of biologists have speculated about these dimensions, among them my friend the very eminent reproductive biologist Professor Roger Short. Various hypotheses exist; the long penis may be more attractive to women or it may be more threatening to competing males. I can certainly believe the latter to be true. As a young teenager, I remember being subjected to the sight of a sixteen-year-old adolescent who, in what was an undoubtedly aggressive display, exposed himself to me, fully erect, in the changing room of the local municipal swimming baths. What is interesting, in retrospect, is that this seemed at the time to be threatening behaviour rather than any mere show of potential sexual prowess. So, just possibly, a long, prominent penis frightens off other males who might compete for the same delights.

Lynn Margulis, Professor in the Department of Biology at the University of Massachusetts, Amherst, and her son, science writer Dorion Sagan, are a very well-known team in the biological sciences. In their book *Mystery Dance: On the Evolution of Sexuality*, they argue that a long penis might have been selected by evolution. They suggest that when a female mated with several males, the male who delivered his sperm closest to the uterine cervix – and therefore the site of fertilization, the Fallopian tubes – would have had the best chance of impregnating his partner and producing offspring.

Roger Short has argued, I think more cogently, that a thick penis (a notable feature in humans) may also be the result of natural selection. He points out that thickness is more likely to produce female sexual satisfaction, and that in consequence a male with a thicker penis would be at an advantage when it came to mating. (Incidentally, in comments reported elsewhere, Professor Short points out that having a penis and testicles is not entirely a bed of roses. Castrates live longer, as several studies show. Testosterone production continues at some cost to the host: for example, heart attacks are more common in men producing high levels of the hormone. For the fertile male, says Roger Short, 'It's a short life but a merry one.')

It is also fair to emphasize that our sexual habits and evolutionary strategies have changed somewhat over the past five or six million years. In Lucy's time (remember, Lucy was the petite *Australopithecus afarensis* female living around three million years ago), the difference in size between men and women was far greater. If we use size alone as an indicator, it appears that Lucy and her contemporaries were more likely to live according to the harem system.

But for modern *Homo sapiens*, there are two opposing forces: the attractions of the harem and the negative social and emotional aspects of promiscuity. The push and pull of sexual instinct might well have met somewhere in the middle. Perhaps monogamy is our 'natural' state of rest, the solution to the complex mating equation. But before all the incurable romantics and the deeply religious get too excited, this doesn't necessarily entail lifelong monogamy. As we shall see, it may mean we are simply built for one sexual partner at a time – with lots of partners over the course of a lifetime.

Till death us do part

The harem system is dangerous, particularly for males low in the pecking order. To be an undesirable, low-status male in a harem-led society is an unenviable fate (though possibly the fate of the male eunuch who traditionally looked after the ladies of the Eastern harem was even less desirable). In the main, our instincts for competition mean that males excluded from the system do not take their fate quietly and are likely to be a source of aggression and violence. This is why polygyny has tended to crop up in societies where power is most centralized; high status enables the Mao Tse-tungs of this world to control and repress any violent uprisings from the dispossessed males.

The harem system is also bound to cause problems among the wives. In Mormon families in which polygyny is still practised – although it is illegal and Mormons have officially renounced it, according to some reports there may still be as many as eighty thousand polygynists in the American state of Utah alone – wives have naturally reported feelings of jealousy and stress which emerge from conflicts between the women. In the USA, increasing numbers of conflicts are also reported by the courts. I do not know how typical is the case of Tom Green, a fundamentalist Mormon from Utah with five wives. He recently faced up to twenty-five years in prison on four counts of bigamy. He took his 'head wife' in 1986 when she was thirteen, and as a result he is also facing statutory rape charges. He had a job, but it did not pay enough to support his twenty-five living children, with another three on the way, so he has also been involved in a long-running welfare scam which allegedly netted fifty-four thousand dollars over a four-year period.

Our sexual psychology is rife with complications. The relative sizes of men and women suggest promiscuity, harems and polygamy, but there are excellent reasons for supposing that monogamy, or at least 'serial' monogamy, has always been the most popular family set-up. Practically, and in evolutionary terms, we can see this makes sense. On the savannah, when conditions were hostile, a male human or pre-human hominid might have found it difficult to maintain and protect more than one sexual partner and their respective children. Rarely might he have been in a position to commandeer enough resources to care for and supply a large group of females, as well as protect them from being snatched by rival males disgruntled at the lack of available sexual partners. To me, it seems quite probable that it wasn't until humans constructed dwellings, with some sort of semi-permanence, that this pattern changed. So the picture of the biblical forefathers, such as Jacob, being able to support several wives and partners is a reflection of increasing stability in human society as we slowly ceased to be merely hunter-gatherers.

Where does plain, ordinary, monogamy fit in? Famously, birds do it, if not bees. Most bird species pair up to breed, both the mother and the father becoming involved in raising the chicks. Baby chicks are not dependent on the mother's breast for their daily feeds, so male birds can take at least half the workload, unlike mammals. Most of the world's mammal species do not pair off into couples; the amount that do is thought to be around only 5 per cent. Most are solitary beasts, and the brief moment of pleasure is more or less the only contact a male has with the mother of its offspring. Some species, like wolves or, as we have seen, chimpanzees, operate on a communal basis whereby a group takes care of the babies born within it, but the males do not seem to know or care which of the babies biologically belong to them.

Though monogamy is relatively uncommon in the animal kingdom, and rarely do the parents split the workload of raising their offspring, there are a few exceptions. Gibbons are monogamous for life, as are a few species of European and North American birds. And so are we, at least in theory; 'till death us do part' is still the defining moment in the modern marriage ceremony. But as we'll see, our evolutionary past, our instinct for survival and even our biology shape our sexual morals too.

Cheating hearts

Even allowing for our high divorce rate, the Western world has been one of the most staunchly monogamous societies in history. 'We are firmly attached', said Saki (the Edwardian author H. H. Munro), 'to the Western custom of one wife and hardly any mistresses.'

Monogamy does not mean, however, that we do not have affairs, because we do. Some estimates of the number of British married men or women who have extra-marital affairs can run as high as 50 per cent. We already know some of the genetic reasons why, it seems, men want to have affairs. They're programmed for sexual variety, to spread their genes; if there's a chance of impregnating another female – especially one who is already married and would therefore not have to be cared and provided for – then he may well have stumbled across the best evolutionary knock-down cut-price bargain of all time: all the benefits of the genetic legacy with none of the work or paternal investment in bringing up the child.

Alongside the chance of hitting the genetic jackpot is the rich cultural tradition of adultery and the pleasures one finds in illicit sex. The joy of adultery is even given a cryptic reference in the Bible: Proverbs tells us that 'stolen waters are sweet, and bread eaten in secret is pleasant'. Extra-marital affairs are, in many hunter-gatherer societies today, as well as in more developed cultures, a rich source of gossip and an important aspect of communal life.

While it's easy to see what a man may gain from 'a bit on the side', what, from the female perspective, is the evolutionary advantage to taking on a lover? We all know that Charles Darwin discovered the idea of sexual selection – the theory that some individuals are inherently more successful than others when it comes to competing for mates – but it seems he was rather innocent about sex itself. He made the assumption that the females of most species were naturally monogamous. It is also true that many of his contemporaries thought that women did not really enjoy sex, and they doubted the existence of the female orgasm. One notable Victorian doctor who specialized in giving sexual advice, Dr Acton, writing in 1860 in his book *Disorders*, regarded it as fact that female pleasure was an aberration and a rarity. And for decades afterwards, many biologists clung to this notion of natural female monogamy.

But as Victorian attitudes died, the reality was slowly revealed. Females of numerous species, not to mention women themselves, knew all about affairs. We now know that for women in all societies infidelity is a regular occurrence. Indeed, genetic studies in certain rural parts of the UK have thrown up the possibility that up to 15 per cent of children were not fathered by their 'official' father. But to be fair, these studies are based only on blood group evidence and possibly deserve re-evaluation.

One way of explaining female infidelity is that it's a woman's way of hedging her bets. The anthropologist Marjorie Shostak describes how Nisa, a !Kung woman from the Kalahari Desert, regularly sleeps with other men, despite being married. 'A woman should have lovers wherever she goes,' Nisa told her. 'If she goes somewhere to visit and is alone, then someone there will give her beads, someone else will give her meat, and someone else will give her other food. When she returns to the village, she will have been well taken care of.' Even in studies of more contemporary societies, this shines through. In the words of one woman, 'Men are like soup – you always want to have one on the back burner just in case.'

The security of knowing there is more than one 'provider' for you and your children is not to be taken lightly, especially in a harsh desert environment where scarcity of resources is an ever-present and dangerous threat. In addition, a lover provides an extra insurance policy: if your husband dies or is killed on the hunting grounds, there is someone else to help you take care of the kids. So, from the point of view of material subsistence and wealth, lovers can be profitable. Nisa could also be lucky and snag a high-status male as an occasional lover, which could bring added benefits and respect from others within the group.

Adultery, for a woman, is also about dipping into the genetic pool. Your current husband may be infertile, or may simply carry poor genes. He may be weak, sickly or a poor hunter. Illicit adultery is one way of introducing different DNA into the litter without destroying the stability of the family structure; and the more desirable the man, the more desirable his genes, and vice versa.

The ubiquity of female adultery makes me wonder about the evolutionary psychologist's portrait of the sexually coy, choosy female. The desire for sex is ever present. In societies with very relaxed morals about non-marital sex, the women are considered just as promiscuous as the men.

The battle's on

Biology also reveals fascinating insights into the long history of female infidelity. In the late 1960s, the entomologist Geoffrey Parker, now at the University of Liverpool, was sitting down to another long day of heady investigation into the behaviour of flies cavorting around dung pats when he made a startling observation. He noticed that the female flies regularly mated with more than one male. Subsequently, he was the first to recognize that sexual selection doesn't actually stop at sex, it can continue even after copulation and right up to the point when fertilization occurs. This triggered off a whole field of investigation into what is now known as 'sperm competition'.

Animal studies revealed that this type of competition is possibly a selective force in evolution, leading to a variety of behavioural, physiological and anatomical adaptations, each aiming to ensure the success of the individual male's sperm. As we've seen, increased testicle size meant one male could flood a female with more sperm than another, increasing his genes' chances of making it across the finishing line and to the egg first. The very fact that this adaptive mechanism has evolved confirms that female animals have not been monogamous over evolutionary history. Had they been, sperm competition in all its guises could not, by definition, exist.

Intriguing evidence of sperm competition and selection is

certainly found in animal and insect studies. One of the most studied insects is the fruit-fly *Drosophila*. It may seem bizarre that a fly has such a major place in the annals of biology, but because of its large chromosomes, its breeding behaviour and the fact that it has many genes known to be very similar to those found in mammals, it has been of great interest to geneticists for a long time.

A recent study by Dr A. G. Clark in the USA shows that among female fruit-flies in the wild, 80 per cent clearly harbour sperm from more than one male, and that, curiously, the male who mated last is the insect who becomes the proud father. In some other species, too, the females, it seems, get a chance to be choosy about the father of their offspring, even as his sperm is on the way to her egg. Sometimes known as 'cryptic female choice', it seems there are dozens of ways a female may control who gets to be daddy after mating with more than one male has taken place. Female hens tend to fancy socially dominant males, but because all males are significantly bigger than the females, subordinate males often manage to mate with females anyway. A recent study showed that hens in a free-ranging wild population frequently ejected semen immediately after mating, and were more likely to do so after having copulated with a subordinate male. There's also evidence of other female animals retaining the sperm of different males in different internal stores and choosing to release it selectively. While all these are observations from animal studies, it makes you wonder how much (unconscious) control human females may have in this extraordinary realm.

But what about human sperm? One survey carried out on a large group of women uncovered the fact that one in every thousand copulations was with a second male. According to the figures and calculations, this is eight times the frequency needed for full-scale, all-out sperm war to have evolved. Though I haven't measured it myself, it's been calculated that a man's sperm volume relative to his body weight is in fact twice that found in primate species which are known to be monogamous. It seems that over time human females have indeed been doing their share of monkeying about too.

Long ago, I came across an extraordinary clinical situation where human sperm competition did not seem to work; and since that case, I have come across other rather similar stories at my infertility clinic.

Margaret B came for investigation of her infertility in her early thirties; IVF was not an option she wanted to consider. Exhaustive investigations failed to find the slightest thing wrong. She was fit, she ovulated, her uterus and tubes were normal, her hormones were fine, she had no immune problems and I could find sperm in her uterus on examination many hours after intercourse. Her husband, too, seemed in good health with an apparently excellent sperm count.

One day, some eight years after she had first come to me with the problem, and following another failed cycle of treatment when I had stimulated her ovaries and performed repeated insemination with her husband's sperm, I said that perhaps the problem could just be with her husband. There was, I said, a small group of men with apparently normal sperm counts whose spermatozoa were incapable of fertilizing an egg. She looked at me for a long time, started crying and said, 'No, it must be me.'

Eventually, her story poured out. She had been sleeping very regularly with her husband, but for the last six years she had been having very regular intercourse, sometimes on the same day and even when she was being treated by me, with her longstanding lover. 'And he has three children, so I know he's fertile,' she told me. There was little I could say beyond offering her comforting noises about the possibility of her resolving which of her two partners she would eventually prefer to stay with, and gently suggested a temporary break in treatment might not be a bad idea.

Three months later, Margaret came to my clinic to tell me that she and her lover had taken a momentous decision. She had just seen him off at London airport – he had decided to leave England to settle in Canada. I arranged to see her again in two months' time for another treatment, but before that, only five weeks after her final farewell at the airport, she phoned to say that she had just missed her period and the pregnancy test was positive. And this time, there was only one possible father.

Much publicity in the press has been given to human sperm competition. Most of it appears to be nonsense, but the stories persist, perhaps because the thought arouses other primitive instincts in those who buy newspapers. Even much of the academic data needs to be interpreted with care. At the University of Manchester, Robin Baker and Mark Bellis have argued that human spermatozoa come in different shapes and sizes because of this possible battle against a competing male's sex cells. They claim that the most common sperm are the standard-issue 'egg-getters' with conical heads and long tails which are designed to swim for their lives. According to their studies, a different type of sperm is also ejaculated. They have coiled tails, so swimming certainly isn't their forte; instead, they act as kamikaze sperm, wrapping themselves around foreign egg-getters thus hampering their progress to the all-important egg. These researchers are convinced that 'sperm competition has been the main force to shape the genetic programme that drives human sexuality'. But I believe their views are fanciful; most of the unusual-looking sperm in human ejaculates are simply abnormal. They merely reflect the fact that humans produce many sperm that are incapable of fertilizing an egg because so many human sperm cells are improperly or incompletely formed, and therefore abnormal genetically or physically.

Whatever the truth of all this, it is possible that adultery is an evolutionary adaptation that has grown up alongside monogamy and long-term commitment. Adultery, especially for women, is a risk; the bloodshed and violence that comes from male jealousy would be disastrous for her and her children. It needs to be balanced by the long-term needs of a protective and helpful mate. The two forces are in the ultimate evolutionary tug-of-war. From our earliest ancestors onwards, the sex lives of humans were never going to be straightforward.

The biology of fidelity

Let's continue to examine the biological roots of our sexual inheritance. If early men and women did indeed have affairs and continually trod the difficult line between monogamy, with its much-needed bonus of security and adultery, with its potential advantage of improving the spread of their genes, what have been the means by which we've inherited these tendencies?

Can we find any evidence in our genes? The reality is that multiple genes control every single physiological process in our bodies, so we should certainly take care when seeking out single genes 'for' extremely complex behaviours. There is the vast impact of the social environment to consider as well. But that said, it's *possible* that in some behaviours, just as in some physical processes, there are master genes much in charge of controlling the whole matter in question.

Tom Insel and Larry Young at Emory University have been doing some interesting work on the genetics of attachment and pair bonding among mice and prairie voles. Prairie voles are endearingly monogamous. After mating with their partner, the male forms a strong social bond with the female and spends more than 50 per cent of his time huddled up with her. They nest together, and the male turns out to be an exceptionally faithful partner and attentive father. Males spend as much time with their offspring as the females, and they're also staunch defenders of the nest and their female partners, staying with them through each subsequent litter and usually for the rest of their lives. In contrast, the male mice they studied are out-and-out cads. They normally abandon the female instantly after mating and have no role in raising the offspring whatsoever.

Insel and Young looked at the brain chemistry of these two species; in particular, they measured the levels of the hormones oxytocin and vasopressin, which are known to be major contributing factors to pair bonding and attachment. When they compared the receptors where these two hormones have their effect in the two rodents' brains, there was a difference. The prairie voles didn't have higher levels of these hormones, but the receptors for them were laid out in a very different pattern from that found in the mouse brain. The researchers went a step further. They focused on and identified the key gene in the prairie vole which appears to control the hormone vasopressin; then, they injected this 'vole' gene into several embryos taken from their mice. These embryos were put into the womb of a female mouse and, eventually, they became successfully pregnant. Male mice born became normal adults in every respect except they had the vole pattern of brain receptors. Most impressive of all was the fact that they were unusually sociable and attentive to their mates, with only the very occasional lapse for the odd 'extra-marital affair'.

Does this mean that each of us has a gene for monogamy or adultery? We have to be careful not to jump to too many conclusions. It is true that we have genes for oxytocin and vasopressin receptors in our brains, but it is also clear that a multitude of genes and environmental factors are undoubtedly involved in the evolution of behaviour. What is important about this kind of study is that it throws light on some of the fundamental links between our genes, the chemistry of our brains and complex social behaviour patterns we have inherited from our early human ancestors millions of years ago.

The love hormone

Whether or not there is a master gene inside us which guides how we bond with our partners is unknown; no-one has seriously looked for it as far as I know. We do know, however, that a closely related gene to that seen in voles, one which codes for the vasopressin-2 receptor in our own brains, varies from person to person. There's also evidence that the action of the hormone oxytocin may have an important part to play in the way we form attachments to the opposite sex.

Have you ever wondered why after sex men are so happy to just roll over and go to sleep? Perhaps it is just the inevitable result of excessive exercise, but possibly oxytocin is the culprit. This hormone plays a number of roles in the body, including triggering uterine contractions during labour and helping a mother bond with her newborn baby. However, it's also released during sexual arousal in both men and women. Levels rise during touching and cuddling, the extra oxytocin making you increasingly sexually receptive. In both sexes, the levels of the hormone peak at orgasm. But that's where the similarity ends, because it seems that the combination of other hormones with oxytocin results in the age-old battle of the sexes running rife at this point. It's believed that oestrogen somehow amplifies the effect of the oxytocin in women after orgasm, so most women become very affectionate, want to cuddle and feel that strong emotional bond with their partner. Men, on the other hand, mostly just want to snooze, and that may be because testosterone inhibits the bonding effects of oxytocin after orgasm.

This intriguing peek into the world of hormonal influences on our sexual behaviour reflects the importance of our biology in reinforcing the way in which women have evolved to become more easily emotionally attached to their partners. As we saw in the last chapter, women tend to find the idea of their partners becoming involved emotionally with other women much more abhorrent than their simply having purely sexual encounters with no strings attached, whereas for men the complete reverse is true. Men's deepest fear is to find they have been cuckolded and tricked into raising another man's child, whereas women don't want to lose out on benefiting from their partner's resources and that all-important security and protection when raising their children. It's both fascinating and remarkable to find that these ancient differences between the sexes when it comes to fidelity are reflected in our very own biology. That should remind us that although our modern culture, society and its traditions have so much sway on our morals, we are also talking about behaviour that has been rooted deep within us for millions of years, and with good reason.

What about starting afresh?

Some species of bird such as eastern robins, as well as other animals such as foxes, pair up for just one breeding season. They are serially monogamous, spending a pre-determined length of time with their mates and helping to raise their young before splitting up. Robins will pair up in the spring and raise their young during the summer months. But when, in late summer, the chicks have flown the nest, the two parents amicably go their separate ways and join a flock. Similarly, a pair of foxes, having produced a litter together, will hunt for food and protect their helpless kits for the summer, and then, without ceremony, they will split up and leave their growing progeny to fend for themselves.

The evolutionary psychologist Helen Fisher suggests that humans pursue a similar strategy. We are designed, she submits, to be monogamous only for as long as it takes to raise a single child through infancy. Perhaps this is reflected in the amount of time our brains are receptive to those initial high levels of the chemical PEA (phenylethylamine), which, as we saw in the last chapter, launch us into those heady eighteen months to three years when we first meet a new partner and fall in love.

In the US and UK the figures are only too clear. Between 40 and 50 per cent of marriages end in divorce. But when is this separation most likely? Fisher has collected marriage and divorce statistics from almost sixty different countries. She has found a worldwide pattern that shows that divorce peaks at around four years into a marriage and then declines. It is a pattern that does not seem to alter despite different cultural norms, marriage practices, divorce procedures or relationship difficulties. According to this theory, every marriage is a divorce waiting to happen; and some understanding of the reasons for this may be found by studying the habits of the hunter-gatherers.

In many traditional societies such as the !Kung, there is a much longer period of breast-feeding than in the West. Breast-feeding is sustained until the child is as old as three or four. !Kung women keep their infants by their side day and night. They will use breast-feeding in the way parents in the West use a dummy or pacifier. This is probably the mechanism by which !Kung mothers space out their pregnancies. Although a woman can occasionally fall pregnant during the period when she is breast-feeding, during lactation ovulation tends to be suppressed by high levels of the milk hormone, prolactin. Births among the !Kung women are about four years apart. The same is true for several other hunter-gatherer cultures such as Australian Aborigines, Netsilik Eskimos and the Dani of New Guinea (although there are other contributory factors, such as their low-fat diets and their active lifestyles, that also have a bearing on post-natal fertility). We should not necessarily think that all human evolution can be deduced from these contemporary traditional cultures, even if they are apparently the closest modern-day comparisons to ancient human culture on the savannah. But there is collusion between cultural and biological factors to space out births over the years – which, after all, has the benefit of not overstretching resources and reducing the group's vulnerability to scarcity or predators.

What are the reasons that force our marriages to break down? Is there any common cause? Interestingly, adultery is the most frequent reason, particularly adultery by the wife, and infertility is second. These two situations underline the fundamental point of human pair-bonding: we enter long-term monogamous relationships in order to have children, preferably our own biological children; anything that compromises this ambition is bound to be destructive in a marriage. As we've seen, jealousy is at its most powerful when a woman has sex with another man. The divisive aspects of infertility are self-evident; the fact is, even the most deeply felt love and respect for one's wife or husband does not necessarily outweigh the instinctively felt need to produce one's own children. And from my own disturbing experience of running an infertility clinic, childlessness is a very frequent cause of depression and severe marital disharmony. Divorce is particularly common among infertile couples.

Very few known societies have banned divorce. The Incas did not tolerate it, nor does the Catholic Church, but in the majority of cultures, both traditional and modern, divorce is commonplace. Couples in the modern Western world are starting to move away from the powerful Christian prescriptions that attempted to protect the sanctity of the lifelong marriage bond; 'no-fault' divorces and annulments are increasingly common methods of dissolving the marriage bond, and cultural squeamishness when it comes to divorce is slowly ebbing away. Some people welcome this trend; as George Burns once said, happiness is having a large, loving, caring, close-knit family in another city.

Whatever the individual reason for separation, we know that relationships do often break down, and the four-year mark is a statistical watershed, even when there are children in the mix. There is no instinctual force which keeps couples together for a lifetime; 'till death us do part' is more often than not a hope beyond hope, a tradition which from an evolutionary perspective has little relevance to the realities of human relationships. It seems that while we're socially monogamous, the truth is, by nature, we're sexually polygynous.

Harems into trophies

There's a bad locker-room joke that says when a man's wife reaches the age of forty he should be able to change her, like a bank note, for two twenties.

The reason you may find this offensive is because it chimes well with the reality of men's sexual preferences. Men like younger women because they can have more babies. Although consciously the vast majority of modern men probably don't see their love of younger women as being for the purpose of having babies at all, most studies suggest it is a powerful factor in modern mating patterns. For some men, the loss of sexual interest that corresponds with the ageing of one's wife can be enough to fatally harm a twenty-year marriage, despite the bonds of love and companionship that have formed over the years. Cue the proverbial male 'mid-life crisis'.

So what happens next? A man who is sufficiently attractive, or who has reached a position of high status or cachet, often ends up marrying a younger woman, and sometimes a succession of younger women. The phrase 'trophy wife' suggests the young, glamorous, sexy new woman in a man's life is for the purposes of ego only, a symbolic catch to reinforce one's status and virility – a bit like wearing the latest, biggest Rolex on your arm, just prettier. It seems the ageing cad is shouting out to the world, 'OK, I may be older, but I'm so powerful that I can snag this pretty young specimen.' Despite the attraction of this concept, it turns out that the 'trophy wife' phenomenon is far more deeply rooted than that, and it has real, practical implications. Men who marry a series of younger women – perhaps a bit like the Donald Trumps and Hugh Hefners of this world – are catching these women in their fertile prime; on an unconscious, biological level they may be striving to maximize their genetic legacy (and at the same time happily reminding everyone of their alpha maleness).

In fact, this small minority of men are practising a form of polygyny. Although they don't keep more than one wife at a time, they're marrying women in their prime and then discarding them, so the principle of polygyny holds. In many cases they will also have to support their wives and their ex-wives, just as in a polygynous harem. And, as in the harem system, they are excluding other men from the chance of mating with these young, fertile women.

So we are back to the *sine qua non* of human sexual activity – procreation. Whether or not we like the idea, we can nearly always trace the permutations and complexities of our relationships, our affairs and our sexual preferences back to the reason why we have sex in the first place – babies. This instinct really does define and shape our everyday lives.

The power of kin

There are several places in the Bible where the need to respect one's parents is emphasized, but it is equally striking that nowhere in the Old Testament, which provides so much of the basis for Western views of morality, is there any suggestion of a requirement to respect one's children. Almost certainly this is because it is automatic human behaviour to nurture one's offspring, and this instinct means that any dictum of this kind is unnecessary. In my view, the worst bereavement any human can face is the loss of offspring, no matter what age. Our society may occasionally trivialize infant death and miscarriage, but I consider from personal clinical experience of many patients that a great number of women never fully recover from this kind of death in the family. Babies are at the heart of our family relationships, networks of kith and kin that go back hundreds of thousands, probably millions of years. They are also the focus of intense love and care among adults. Feelings of love for our children seem to be ingrained in our subconscious; babies do not just get attached to their parents, their parents also get attached to them.

The loving relationship between parent and child is, on the face of it, an obvious biological adaptation. Those children who are doted on and cared for are more likely to survive the ravages of childhood and last until they are sexually mature and able to reproduce themselves. Love, in this sense, follows the genetic pathway; our genes survive through our children, and so on through our children's children.

In the rough and tumble of the natural world this fact has some disturbing consequences. Take the example of a pride of lions. Most prides are made up of juvenile cubs and adult females, all of whom are related – as sisters, mothers, aunts and so on. Alongside the female clan are a few non-related adult males; they will have ousted the previous group of males, who would have been too old or too weak to protect their place in the pride. When they mature, the young males desert the pride and go hunting for an adoptive family.

Lionesses nurse their cubs for a year and a half. During that time they cannot become pregnant and their fertility returns only once the cub is weaned. When a new group of males ousts the previous incumbents, it's likely that one or more of the females will be nursing their young and thus unavailable for further reproduction. The new group is aware that the existing cubs were fathered by members of the previous group. On arrival, the new males swiftly and brutally wipe the slate clean; they seek out and murder every single cub in the pride. The females are once more free to mate and reproduce, and this time the new males will be the genitors.

The brutality and mechanical violence of animal behaviour still has the capacity to shock us. Yet we can find the same kind of behaviour in humans, if we look carefully enough.

The Ache Indians, a small hunter-gatherer society from the interior Atlantic forests of northern Paraguay, have an unusual tradition that mirrors the killing of the lion cubs. When a father dies, the villagers ceremonially kill one or more of the man's children, even if the mother is still living. In the eyes of the tribe, the children are sacrificed to appease the gods, somewhat in a tradition shared by a number of human societies from the followers of Moloch, reviled by the ancient Hebrews, to the Aztecs. But what seems to us incredibly inhumane is in fact a practice reminding us of one of the fundamentals of evolution, which may well share its roots with the ruthlessness of the lions' genetic cull. The Ache widow will remarry within the village, and it is not in the new husband's genetic interests to support the dead man's children. The child sacrifice may be a vestige of this brutal calculation, a symbolic gesture to maintaining the new husband's genetic fitness.

In the case of the lions, however, biology is most definitely king, and we can be almost certain that their actions are a product of evolutionary adaptation. But why would lions have evolved such destructive behaviour? They are large mammals, with a long and slow reproductive cycle. It's unlikely their numbers would ever have grown to the point where this kind of killing of cubs wouldn't have damaged the success of the species. Their behaviour seems, from the point of successful maintenance of the species, completely deranged.

Group success?

The answer to this conundrum lies in a truth the realization of which represented a seismic shift in our understanding of evolution and natural selection.

For years, evolutionary biologists built their theories on a largely faulty assumption. They believed, as had been originally suggested by Charles Darwin, that evolution would act *on an entire species*. This means that if a physical or behavioural trait, like bats' ability to use ultrasound for navigation or the development of more powerful back legs in a frog, was of benefit for the species as a whole, then it would be a perfect candidate for natural selection. Evolution, from this point of view, had the interests of the entire species at heart. This theory of group selection was rife in the 1950s and early 1960s and seemed to explain many animal behaviours such as those seen in the lions.

Imagine a species of fish-eating birds that live around the shores of an inland lake. The birds have evolved to dive into the water and spear the fish with their beaks. But they are hampered by poor eyesight; they can only see the biggest fish, so their catching ability and therefore their population is limited. Say there is a mutation that brings an advantage, such as sharper eyesight. Now, the birds that have this mutation are able to spear the smaller fish too and because there are so many more fish to choose from, they thrive. The mutated birds are so successful that the mutation spreads quickly throughout the entire population. But there are only so many fish in this particular lake, and soon the supplies start to run dry. The birds are growing too fast, reproducing too quickly and too many young fish are eaten before they themselves have a chance to reproduce.

The 'group selection' camp of biologists would have said that the birds were not necessarily doomed. They might have argued that the species could evolve into a population that takes better care of its resources. In the long term, given that there are no other lakes nearby, there is no other option. Species can evolve to suit the limitations of their environment, which is why we see such precisely balanced eco-systems in which every species plays its allotted role.

They would, however, almost certainly have been wrong. In this case, the birds with the sharp eyesight should, individually, *always* do better within the group. There is no possibility that the 'poor eyesight' gene could spread throughout the population, even if the birds with poor eyesight are, in the long term, a much better bet for the species as a whole. Our flock of fish-eating birds is destined for evolution's dustbin. The likely truth is that the individual will always strive for its evolutionary success over and above the group as a whole.

The idea of group selection was rapidly dropped by most scientists when it became clear that only in certain very rare circumstances would the group be the unit of selection. One area where things were much clearer as a result was sex selection and the ratio between males and females.

Is it a boy or a girl?

Just imagine if management consultants were employed to make human reproduction more efficient. They would quickly realize the human race would be far more productive if its population were made up of 10 per cent males and 90 per cent females. Even a male population as low as 1 per cent would be enough to fertilize the female majority on a regular basis. Instead of men lazing around having clocked off at the mating factory, every member of the species would be employed doing what they do best: making babies, in the most cost-effective, downsized manner possible. This harem system is an innovation of which Henry Ford would have been proud. So, if evolution is, as we know, nature's very own management consultant, why on earth is 50 per cent of the human population male?

There is a practical consideration: despite the reductionist way we've been thinking about mating rituals, reproduction was certainly not the only activity on the savannah. Our hominid ancestors had to eat and fend off predators and scavengers, and the male members of a group must surely have helped significantly in this respect, although no-one can know for sure the extent of their contribution to the communal larder. But even if men were entirely useless in those respects, it turns out that there'd still be a sex ratio of 50:50 in the human population. The reason is that, as we have seen, natural selection does not appear to operate on groups, just individuals. A particular mutation or genetic trait has an evolutionary benefit only if it means that the chances of one single organism's genes are passed on more effectively, or in greater numbers, to the next generation.

If a species is completely monogamous, then the sex ratio is self-evident; a male child has as much chance of siring offspring as a female child. But what if a species is polygynous? If *Homo sapiens*, as a species, was slightly polygynous in the past (males commandeering more than one female), would it not be a good strategy to give birth to more females, who are pretty much guaranteed always to have children, rather than males, who may sometimes be completely excluded from the mating process?

You may have already guessed the answer. Every human child has a biological mother and a biological father. Even if one male controls a harem of ten females (thus excluding nine other males), that one male will sire ten times the number of children. In a system like this, the odds are against most males reproducing, but when one male does, he hits the jackpot. So producing males is, in the long run, just as profitable as producing females.

From the point of view of evolution, we all play the long game. When mathematics is applied to the problem, modelling these different reproductive tactics over many generations, the reason why we are as we are becomes clear: a 50:50 sex ratio is a state of equilibrium. If more females than males are present in a population, then the best tactic for maximum genetic payoff is to produce more males.

Natural selection, acting on individuals, will even out the ratio in the long run.

Looking after your own genes

In the early 1960s, an unknown American graduate student called William Hamilton was becoming frustrated with traditional evolutionary biology, which tended to be rife with what he felt were mistakes about the notion of group selection. He attended lectures given by old-time biologists who believed firmly that that was the primary mode of evolution. After attending one of these lectures at the University of Chicago, Hamilton left, muttering under his breath, 'Something must be done.'

In early 1963, Hamilton published a paper in a somewhat obscure biological journal. It was called 'The Genetical Evolution of Social Behaviour', a coldly mathematical analysis which outlined one of the most startling revelations in evolutionary biology. It began, 'A gene is being favoured if the aggregate of its replicas forms an increasing proportion of the total gene pool.' As Andrew Brown, author of *The Darwin Wars*, points out, Hamilton's paper has the quality of a Turkish bath: one bathes in the warmth of his preface before plunging into the icy waters of the mathematics which make up the bulk of the paper. The mathematics, though, contain an inescapable conclusion: evolution acts on the gene, not on the level of the species, the group, or even the individual organism. This means that natural selection actually selects successful genes rather than successful people, and those genes will spread throughout the population. Hamilton realised that this truth has some extremely interesting implications.

If a gene is to be favoured by natural selection, if it is to survive and not die out, then it needs to enhance the survival and reproductive chances of the organism in which it exists. A gene that boosts the immune system and allows better resistance to disease is likely to grant a longer-than-average life to, and therefore a better-than-average number of offspring for, its animal host. Reproductive success means that the gene will be spread more quickly through the population than its predecessor, a less useful gene that leaves the animal host more susceptible to disease. This early version of the gene will eventually die out if the new gene is allowed to spread unhindered through the species. For any particular gene to survive, it must be dispersed through the population and the species. Its DNA must code for a physical or mental trait that increases the reproductive success of an individual animal. Even a gene that gives the animal a tiny percentage advantage in the resistance to disease could, over many hundreds or thousands of generations, find its way into each and every one of the genomes across the entire species. We can all trace our ancestry back to those with the successful genes; the rest were doomed to end at an evolutionary cul-de-sac, the dead branches of humanity's family tree.

This has an all-important effect on our individual behaviour towards our kin. Our children carry our genes; in fact, we share 50 per cent of our genes with them. We also share 50 per cent of our genes with our siblings, because each brother or sister receives half of his or her genes from either parent. The percentage decreases as our blood relationship becomes more distant. We share one quarter of our genes with half-brothers or -sisters, the same as we do with first cousins. Because our children, siblings and cousins carry a proportion of our genes, it is in our interest to help them survive, prosper and reproduce. It doesn't matter that these genes will find success in another person's body. The fact that our own genes find success, whoever happens to be the host organism, is what is important. From a gene's point of view, any organism that contains copies or partial copies of itself is worth preserving.

Hamilton's breakthrough ideas became known as the theory of 'kin selection'. Natural selection, according to this view, operates simultaneously on an individual and its kin, because all carry a proportion of the same genes. Hamilton's equations lay bare some elegant truths about human and animal behaviour. This is a subject we shall return to, because it underlies a great deal of new thinking about human nature, altruism and our relationships with others.

So blood *is* thicker than water. But as an aside, it is interesting and alarming to see how often and how easily these instincts are used, and sometimes manipulated. I believe the feelings we have towards our family and our group go some way towards explaining xenophobia and racism – or at least, how some political figures exploit these strong feelings. As I write, there is a resurgence, it seems, of right-wing politics endeavouring to appeal to the baser instincts of the electorate. One old slogan of Monsieur Le Pen, the French fascist leader, will perhaps serve to illustrate what I mean: 'I like my daughters better than my cousins, my cousins better than my neighbours, my neighbours better than strangers, and strangers better than enemies . . .'

There is some extremely interesting and to my mind reassuring scientific evidence about racism and our 'racist' instincts. Dr Robert Kurzban, working with John Tooby and Leda Cosmides, evolutionary psychologists at the University of California in Santa Barbara, has confirmed that automatic processes in people's brains compute the race of those they encounter. One possibility is that these processes may be deeply ingrained and that people cannot help categorizing people by race. Given that categorizing people into groups nearly always leads to discrimination, this idea would be very discouraging. However, an evolutionary analysis would suggest a more hopeful view of human nature. During our history on the savannah, our ancestors would have lived in a world where the sex and the age of another person were likely to be very important in the immediate structure of their society. But hunter-gatherers travelled by foot, and residential moves of more than forty miles or so would have been uncommon, which makes it highly unlikely that the average hominid ever encountered people from populations that, genetically speaking, differed substantially from himself. Therefore, they argue, it's unlikely that natural selection has left us with any preferential recognition of race.

Kurzban and his colleagues set out to try to show that, as a result, our attitudes to age and sex – ageism and sexism – are much more deeply ingrained than those of race. The experiment was complex and used extensive statistics, but the essence of it was simple enough. In a variety of experiments, they showed a series of volunteer undergraduates from a variety of racial backgrounds photographs of individuals wearing similar clothing from one of two rival basketball teams, with identically coloured shirts in some experiments and different colours in others. The photograph of each individual 'basketball player' was paired with a verbally given sentence which suggested allegiance with one or other basketball side – for example, 'you were the ones that started the fight'. The volunteers were then asked to sort the basketball players into one of two 'coalitions' or groups. After sorting and analysis of their results, the experimenters found evidence that although recognition of race was soon forgotten by their volunteers, memories of the perceived age or sex of the subjects of the various photographs persisted. They concluded that attitudes towards gender or age are matters of much deeper instinct.

The Cinderella effect

We can be almost sure that the genes carried by one particular male lion haven't evolved by allowing another lion's genes to flourish at their expense. There is no evolutionary sense in expending time and energy looking after cubs which are not your own; in fact, it's an evolutionary disaster of a situation. Besides which we also know that the mathematics does not allow it. Genes that led their carrier to pursue such a hopeless strategy would simply not have been successful.

The harsh realities of the mathematics of evolution have some profound implications for human society. Although it's sometimes difficult for us to think of human behaviours and relationships in this way, studies such as those by Margo Wilson and Martin Daly, the Canadian psychologists, have helped us to see how things work in our own everyday lives. As they say rather self-deprecatingly of themselves, they do field and laboratory work on desert rodents and human homicide which they 'treat as a window on human passions and antagonisms'. Wilson and Daly described a phenomenon they call the 'Cinderella effect' which shows with startling clarity exactly how our family relationships are sculpted by genetic calculation.

Evil stepmothers have been a staple in myths and fables from every conceivable age and corner of the world. At the heart of the Cinderella story is the cruelty the heroine suffers at the hands of her wicked stepmother and stepsisters. In a Chinese version of the story, the Cinderella character, who is named Benizara, is sent with her stepsister to collect chestnuts, but her stepmother has cruelly torn a hole in the bottom of Benizara's bag and when night falls the girl's bag is still empty. Benizara is afraid to return home because she knows she will be punished. But, as in all good fairytales, there's a happy ending around the corner. In the woods, Benizara meets an old woman, her fairy godmother, who gives her a magic box which later will be found to contain a beautiful gown which goes on to perform the same function as the glass slippers. There's also a large bag of chestnuts to take home to the evil stepmother.

So entrenched is the theme of the wicked step-parent in stories around the world that it's not surprising to find the pattern reflected in the grim realities of domestic murder and abuse. Wilson and Daly scrutinized the statistics of 147 cases in Canada between 1974 and 1983 in which children were killed by their parents or step-parents, and their conclusions were clear: very few of them were killed by their biological parents. The vast majority of children killed in these domestic situations are murdered by step-parents or adults who are in a sexual relationship with the child's parent, whether or not they take a parental role in their adoptive family. In fact, their study showed that a child living with at least one non-biological parent was an incredible seventy times more likely to be murdered than a child living with both its biological parents. It is a simple fact of evolutionary biology: parents don't want to waste their precious resources on children who are not their own genetic offspring, nor do they want to share the limited resources they have found for their own offspring with someone else's if they can possibly help it.

The Cinderella effect is not just an American or Western phenomenon. Among the Ache Indians, if the children are not sacrificed after the death of their father, their prospects as stepchildren are not auspicious. Of a group of sixty-seven children raised by a mother and a stepfather, an extraordinary 43 per cent had died by the time of their fifteen birthdays. Of children who were raised by both biological parents, 19 per cent had died by the the same age – still high, but the odds are definitely in their favour.

Daly and Wilson were careful to exclude any factors that could skew the result. They took into account the age of the parent or step-parent, whether or not they had a personality or mental disorder, and their social and economic status. And, rightly, they are quick to point out that the murder of children is, in most societies, a very rare phenomenon. Most step-parents are generally loving, peaceable and devoted. But the statistics reveal the sharp end of the truth about how kin selection really works. Without the genetic bond, the parental bond is bound to be weaker.

These harsh conclusions do not just apply to children and their step-parents. Domestic murder rates in general are excellent indicators of the mechanics of kin selection. Detroit used to have one of the highest murder rates in the Western world. Around a quarter of murders in 1972 took place in a private home or involved solely family members, but only a quarter of these killings involved two people who were biologically related; the rest involved lovers, husbands, wives or step-parents – family members, but not blood relatives.

Unrelated exceptions

As soon as we start to think about the basic mechanics of kin selection in any depth, we can't help but notice that in our everyday human lives there appear to be a number of exceptions. We all know only too well that sometimes people do carry out selfless altruistic acts – directed at people who are not actually biologically related to them. How does this reality sit with the idea that we are lean, mean evolutionarily charged machines destined by our very nature to direct all our energies into furthering the spread and success of our own genes, not anyone else's?

Such acts could range from simply giving a birthday card to the person in the office with whom everyone knows you've never seen eye to eye, to diving head first into a lake to rescue a drowning man you've never met before. The bottom line is we seem to be much nicer to other people than would be expected if this were simply to benefit our selfish genes. This is a theme to which I return at the end of this book, but two examples from the animal world which provide some insights are worth giving here.

When faced with a potential predator, one or two guppy fish usually swim away from the rest of the school towards the looming threat, so that they can inspect it. It seems that the brave scouts deliberately risk being eaten, but they're also doing the rest of the group (most of whom will be unrelated) a favour by checking out whether or not the intruder is actually dangerous. Likewise, a female vampire bat who has recently eaten her dinner of warm blood taken from a sleeping pig will happily regurgitate the entire meal to save an unrelated nest mate from starvation. What on earth is going on? Are these fish and bats true altruists, and if so, how have they survived? Although it may seem that evolution has not made us all into selfish beasts, the truth is probably different. These creatures may be carrying out apparently selfless acts which will benefit other unrelated individuals, but they are also acting on a 'tit-for-tat' basis. The vampire bat will only perform her kind act of charity for another bat with whom she's been a frequent roost mate, and she's more likely to donate blood to a bat which has gone out of its way to help her in the past. And an ingenious experiment using mirrors which tricked a guppy fish into thinking its brave counterpart was either going all-out with him to inspect the predator or turning back at the last minute and leaving him all alone proved the same thing: the fish only behave co-operatively in their risky mission if they know for sure their scouting partner is going to back them up. If they get let down, they won't risk being left in the lurch by the same individual again.

What does this say about humans? Mathematicians and evolutionary biologists believe that reciprocation is indeed a powerful driving force in our choice of actions. Perhaps that birthday card was given because the person in question had made the first move in the conciliation and brought you a cup of tea earlier, or even because the so-called kind act would make other people in the office think differently about you. Besides which, we've all learned, consciously or otherwise, as we grew up in this society that reciprocity leads to social cohesion – getting on better with everyone around you – and ultimately that's got to be good for you and good for your genes. In the world of evolutionary biology, the bottom line is that selfish genes, in their many guises, lead to evolutionary success. Philanthropists be damned!

Busy bees, wasps and ants

Despite our tit-for-tat leanings, in modern Western culture the family is still the main focus of social and personal life. Family relationships – how we act towards family members, how we help them, how we compete with them – are visible clues to age-old instinctual habits, and Hamilton's theory of kin selection is an immensely powerful tool with which to investigate these habits.

As Hamilton's theory goes, the most curious families in the animal kingdom are those of the social insects – bees, wasps and ants. Social insects have an enormous impact on the planet and the lives of other species, mainly because of their vast numbers. Any single hectare in west Africa is populated by an estimated average of twenty million ants. Insect colonies can grow to gargantuan proportions – cities which, measured by population, match the population of the largest human cities of the twenty-first century. The population of Tokyo, currently the largest city in the world, is around twenty-six million; the African driver ant *Dorylus* lives in colonies of up to twenty-two million workers. Their combined mass is more than fifty kilograms, and they feed off and protect a territory of a massive fifty thousand square metres.

It may seem strange to think that the behaviour of insects can tell us anything about ourselves, but their study is absolutely relevant because of the way their selfish genes are passed on to the next generation. The co-operation among these tiny creatures is legendary. They care for their young together; they hunt in crowds and group together to carry food, performing feats that would be beyond any single ant; the colony's soldiers willingly engage potential predators in combat and sacrifice their lives for the good of the nest. Inside the colonies they maintain intricate systems of food distribution. Together they can manage the microclimate inside the nest. *Polygerus* ants in the Chirichaua Mountains in Arizona have even been known to raid nearby nests, kidnap the pupae and return home with their triumphant prizes. The enemy infants are raised as their own offspring and turned into 'slaves' who work 'willingly' for the good of their new hosts.

The family structure of social insects is called eusociality, and it's evolved independently at least a dozen times across the millennia. Eleven eusocial species are in the insect order known as Hymenoptera, which comprises bees, wasps and ants. The other examples include termites and a few odd examples among aphids and beetles.*

The vast majority of eusocial societies are based around female workers; the males simply have to reproduce, to fertilize the queen, an outsized matriarch whose job it is to produce eggs. Workers, soldiers, drones, queens – all have their allotted roles to play. But the oddest thing about eusocial insects is that there is generally a caste of female workers who are sterile. This fact puzzled evolutionary biologists for years. How could natural selection have produced individuals who are completely unable to pass on their genes to the next generation? Why were they so entirely subservient to the cause of the community? In fact, although at first sight these insects appear to be a huge exception to Hamilton's theory of kin selection, the idea actually goes a long way towards explaining it.

In most social insects, there is an unusual genetic difference between males and females. When a queen produces an egg that is unfertilized, it still develops – and always into a male adult. It is one of the examples of successful virgin birth, or parthenogenesis, in the animal kingdom. If the egg is fertilised in the normal way, then the result is a female. This means that males have genes only from the mother; they are what is known as haploid as opposed to diploid, with half the normal complement of genes.

Therefore, the basic rules of reproduction mean that the genetic relationship between sisters, brothers and offspring is significantly altered. If, as is often the case, the queen is fertilized by just one male, then all the daughters she produces will share 50 per cent of their genes with the queen; but they will share 100 per cent of the father's genes, since all of his half (haploid) complement of genes are used

* There is also one mammalian species that has many of the features of eusociality – the African naked mole rat. These small burrowing rodents live in huge cave complexes in groups of several dozen. A large queen sits in a central chamber surrounded by reproductive male drones and sterile female workers; the latter care for the queen and all her progeny. They co-operatively dig for tubers, with 'chain gangs' passing the food along the network of tunnels.

in fertilization. We humans inherit 50 per cent of genes from our fathers and 50 per cent from our mothers, which means we also share 50 per cent of genes with our siblings, but because of the 100 per cent genetic inheritance from their father, ant, bee and wasp females share 75 per cent of genes with their sisters.

Were these females to have daughters themselves, they would share only 50 per cent of genes with their offspring, so from the point of view of gene survival and replication among these insects, helping to raise one's siblings is a far more productive strategy than actually having babies. In technical terms, they 'maximize their inclusive fitness' by helping their siblings; bizarrely and quite uniquely, here sterility is a winning evolutionary strategy. So-called 'honeypot' ants will spend their lives lodged in a tunnel wall, bloated with liquid food which can feed the rest of the nest in time of famine. Others will be assigned the hard manual labour of nest excavation. In some species, soldier ants will enthusiastically conduct kamikaze missions, exploding on impact and killing themselves and their target, a predator and potential threat to the colony.

But while kin selection is a powerful theory, its explanation of the evolution of social insects doesn't reveal the whole story. Consider the fact that if the queen mates with more than one male, the sisters are actually half-sisters, and therefore genetically they should be better off reproducing than helping their siblings survive at the expense of their own lives. Perhaps the queen, through adjusting and tweaking the gestation process, imposes her own evolutionary interests on the behaviour and fertility of her offspring; maybe the sterility of the worker castes came about through this process. This theory suggests that the queen could actually manipulate her offspring to act in ways that aren't to their genetic advantage – a kind of natural selection by proxy. Like the kidnapped Arizona slave ants, the queen's own children could have been duped into maximizing her own genetic fitness. It seems that other types of selection forces might have colluded to produce these extraordinary insect communities and their behavioural patterns.

Infanticide

Evolution is cruel, and it is endlessly creative. Its blindness to morality, pain and suffering has conjured up terrible fates for members of many different species: the lion cubs eaten by the new males of the pride, the kamikaze ants, the male spiders who are eaten mid-coitus, the ant soldiers who commit suicide. All are acting according to the mathematical equations. They are all simply doing their best for their genes.

Humans, too, are capable of deeds that appear brutal and amoral. One of the worst of these misdeeds is infanticide. How could our warm and tender feelings towards the smiling face of a newborn be shoved aside to allow a cold-blooded murder to take place? Is it really possible that the murder of a child is something for which we have an *instinct*? In fact, infanticide has been a feature of practically every known country, traditional or modern.

The Yanomamo, the people who live along the Orinoco River in Venezuela, practise occasional infanticide. In his studies of these people, Napoleon Chagnon reported that if a woman has a second child too soon after the first, she will often be forced to kill the newborn. The wife of a village leader admitted that she had done exactly that, and the reason, she said, was that she needed her breast milk for her two-year-old. The harsh practicalities of the decision did not diminish the grief she felt in having to carry out the killing. Love, even love for one's offspring, does not conquer every other consideration. Whether or not evolution has built us in such a way as to regulate our family population in this manner is a difficult point to prove, especially because there are such vast differences in infanticide rates across different cultures and continents.

Traditional African cultures like the Kipsigis in Tanzania or the Lese in the Ituri forest rarely or never kill their children. However, South American cultures are much more prone to committing infanticide, the Yanomamo being a good example. Some think this is to do with the ecology of their respective environments and the ease with which they can provide for their families, but the argument is less than convincing. If we do have an instinct for infanticide, it seems that it only comes into play when there are powerful reasons

– reasons that are generally a matter of survival, of life and death. Better to feed one child than have both at risk from starvation or malnutrition.

But there is one story of a child murder that goes against the grain of any practical calculation. It stands out as a cautionary tale, a warning not to ignore the startling power of human culture and beliefs over our nevertheless deeply rooted inherited instincts.

The village of Alinagar in the north Indian state of Uttar Pradesh is utterly unremarkable. Perhaps the most remarkable thing about this village is that it's relatively close, about two kilometres away, to a police station, of which there aren't many in this part of India. Basking in the oppressive heat of the dry season, there is a collection of ordinary, small, brick-built houses along rutted streets and an unmetalled town square. The village is surrounded by rich fields, a slow-moving river, buffaloes and plantations mostly filled with the long stems of sugar cane. For most of the long, hot, windless afternoons the villagers doze in the shade of their houses to the sporadic barking of pye-dogs in the background. Not obviously a place to recall the story of Romeo and Juliet.

Everybody in this tiny community knew about the romance between Vishal, a fifteen-year-old boy, and Sonu, a sixteen-year-old girl. Their families, after all, lived less than twenty feet apart. Stories in the Indian newspapers vary about what actually happened, but it seems that in August 2001 a neighbour caught them together as they chatted on the roadside next to a bush. She accused them of having 'suspicious intentions' and dragged them into her shed. And then she summoned their families. The teenagers had not been caught *in flagrante delicto*; they were not even holding hands. Their crime was far more primal and ancient: they were from different castes. Under India's enduring system of social stratification, a relationship between the pair was unthinkable. Sonu was a lower-caste Jat, very much the predominant caste in these parts; Vishal was a Brahmin from the only upper-caste family in the village.

Sonu's mother and father, Surender and Munesh, felt there was only one way to escape the terrible social humiliation their daughter had heaped upon them: they would have to kill her. Some say that they locked the two teenagers in a room, others in the roof of the house, before killing them, but the most credible account is that, aided by three neighbours, Sonu's parents beat both children until they were bleeding profusely. The screams woke various neighbours but in spite of this they proceeded to strangle Sonu in a dark shed, with its abandoned bicycle and mattresses, in front of her terrified boyfriend. After that they got a rope, made a noose and hanged their daughter. Then they called on Vishal's family, demanding that they kill their son. When the boy's family refused, the girl's parents took it upon themselves to hang the boy.

The entire village knew what was going on. According to local reports, villagers flung the teenagers' bodies onto a buffalo cart and hid them under sacking. At three that morning, the villagers walked silently to the local cremation ground, ten minutes away. There, they burned Vishal and Sonu on a joint pyre made from cow dung. Sonu's parents tossed on paraffin for good measure, against all the traditions of Hinduism, so the corpses would burn more quickly. They then surreptitiously threw the ashes into the Katha River.

Luke Harding wrote movingly about Vishal and Sonu in the *Guardian* on 14 August:

> Alinagar is apparently now almost deserted. Most of the villagers have fled for fear of arrest. The buffaloes are unfed; Vishal's house is ransacked and empty. We find only Sonu's sister, Babita, and elderly aunt, Dagiyayi, sitting outside the family home. Neither sees much wrong in Sonu's brutal demise. 'I'm not happy. But Sonu was on the wrong path,' Babita tells me, as she soaps a bucket full of clothes. 'My parents did what they had to do. We were under compulsion.' Did Sonu love her parents? 'Sonu used to love her parents very much,' she says.

The Alinagar case was at least the third such incident clearly documented in this region in ten years, and apparently the police have another forty-seven cases under investigation. The district magistrate, Manoj Singh, was quoted in the Indian newspapers as saying that the case of Vishal and Sonu was in accordance with local tradition: 'There is a theory here in such cases which states: either you kill your son and I'll kill my daughter or you kill my daughter and I'll kill your son.'

It is extraordinary that such a tradition – and a tradition that has come in for vocal criticism within India as well as outside it – can triumph over a parent's instinctive and apparent love for his or her children. The power of a belief system like that of the Indian caste restrictions is extremely deeply rooted and respected, sometimes above all other considerations. It shows us that culture has the capacity to transform or entirely dominate our biologically ingrained instincts, and that we should never underestimate its power.

In all but such tragic cases, however, our genes and their preservation provide us with intensely powerful motivations, and there is no area of human life where they are more deeply felt than in the family. Kin selection is a persuasive theory of how we have come to live lives which are defined by our parents, our children, our brothers and sisters. Polygyny, monogamy, marriage, children – all these relationships are inextricably bound up in our genetic heritage. Shadows of the savannah will always be present, cast over modern mores and ways of life, sometimes with a pleasing match between the way we act and the way we think we should act, but sometimes causing almost unbearable tension. Slowly, in certain fields of study – such as Daly's and Wilson's work on the 'Cinderella effect' – we are starting to grasp the very basic truths about human relationships, and not all of them are easy to accept.

chapter**five**

Risky Business

Survival of the fittest

We usually think of the phrase 'survival of the fittest' as part and parcel of Charles Darwin's theory of natural selection. In fact, it wasn't coined by Darwin at all. It was his colleagues of the time who convinced him that it was a catchy and easily understood way of explaining his ideas to the public. Although the deeply academic Darwin took some persuading, the phrase did eventually become attached to his theory of evolution.

The person responsible for thinking it up in the first place was a contemporary of Darwin's by the name of Herbert Spencer. In 1852, Spencer began to think that contemporary human society was subject to the process of natural selection just like species in the wild. Cells combine to make up organisms, and organisms themselves combine, in some animals, to make up 'superorganisms', or societies. In his view, society and the natural world could be viewed through the same scientific lens: 'Civilization', he said, 'is a part of nature; all of a piece with the development of the embryo or the unfolding of a flower.'

For Darwin, the 'fittest' were those organisms biologically most suited to their environment at any given time. Spencer, extending the model to human civilization, thought the fittest were those people most suited to their society, and for him this meant they had the most highly developed moral, physical and mental qualities. According to this view, individuals within society were like elephant seals or bull elks competing within their herds; and, he said, if natural selection was unfettered the 'weaker' members of society would be eliminated in the competition for resources. Those lacking money, education, strength of will or talent would eventually be outnumbered by more successful and fertile members of society. The laws of nature would see to it, because in Spencer's eyes, biology was destiny.

But Victorian England was not obeying these laws of evolution. The citizens who reproduced most prolifically weren't members of the upper classes – the sons of army officers, lawyers or gentlemen farmers. These members of the so-called moral and intellectual elite were marrying late and having relatively few children; their material, intellectual and physical riches were not spreading throughout the population. Instead, the 'defective' members of the society – the struggling poor, the powerless, the uneducated – were breeding out of all proportion to their 'fitness'. City slums were filled to overflowing with what Spencer saw as the 'unfit'.

According to Spencer, this meant that natural selection had been thrown into reverse, which to his mind amounted to a wholly unnatural state of affairs. He believed that the government should be careful not to oppose the natural biological development of civilization. He railed against any attempts by government or other civic-minded groups to help the poorer classes. Charity, welfare and other philanthropic activities would only make matters worse. He believed that they must let the poor be poor, and then they'd go the way of the dodo. And this wasn't just a social policy, it was morally and biologically right.

Spencer's strident views became a kind of American gospel. His ideas were much in tune with that entrepreneurial society, chiming with the needs of the new American capitalists. After all, the USA was rapidly becoming richer and more self-indulgent than any other nation in history. Now, because of Spencer, no-one needed to feel guilty about his personal good fortune. The rich man was the innocent beneficiary of his own superiority. Personal wealth should be protected, as no government could interfere without inhibiting the process by which the human race was improving.

Even now, how very plausible Spencer sometimes sounds:

> There are many very amiable people who have not the nerve to look at this matter fairly in the face. Disabled as they are by their sympathies with present suffering . . . they pursue a course which is injudicious, and in the end even cruel. We do not consider it true kindness in a mother to gratify her child with sweetmeats that are likely to make it ill. We should think it a very foolish sort of benevolence which led a surgeon to let his patient's disease progress to a fatal issue rather than inflict pain by an operation. Similarly, we must call those spurious philanthropists who, to prevent present misery, would entail greater misery on future generations.

Many Victorians shared the belief that our genetic heritage determined our place in the natural hierarchy of life.* And there were far more noxious proponents of this kind of extreme biological determinism. One culprit was a member of Darwin's own family.

Sir Francis Galton was a cousin of Darwin's. He was a brilliant polymath, an important African explorer, a travel writer and geographer. He was the meteorologist who discovered the anticyclone, he was a pioneer in the use of fingerprints to identify individuals, the inventor of regression and correlation analysis in statistics, and in 1883 the founder of the eugenics movement devoted to the genetic purification of society. It was a scheme designed to solve the very problems that had caused Spencer such anguish. Galton wrote in *Hereditary Genius*:

> I have no patience with the hypothesis occasionally expressed, and often implied, especially in tales written to teach children to be good, that babies

* The virtues of good breeding were also prized in popular Victorian fiction. Oliver Twist, for example, was born in a workhouse and subjected to the harshest and most humiliating childhood imaginable; but Oliver nonetheless grows up to be honest, kind and gentle. The Artful Dodger, on the other hand, a product of the same kind of upbringing, ends up with an ignoble and cunning temperament. Why? Oliver, we later find out, was born of respectable middle-class parents – the underlying message being that his genetic heritage will overcome all the trials and hardships of his desperate situation.

> are born pretty much alike, and that the sole agencies in creating differences between boy and boy, and man and man, are steady application and moral effort. It is in the most unqualified manner that I object to pretensions of natural equality. The experiences of the nursery, the school, the University, and of professional careers, are a chain of proofs to the contrary.

Galton was determined to see the 'fitter' elements of the population reproduce at the expense of the 'unfit'; in his view, the middle and upper classes possessed innate superior abilities which should be passed down the generations. Early marriage and large families should be encouraged among these classes to ensure the survival and expansion of the nation at a time of great upheaval and change across the world. It was a way to perfect the human race by, as he himself put it, 'getting rid of its undesirables whilst multiplying its desirables'. Galton was a patriot and full of good intentions,* but his views were extremely suspect. It is perhaps a wonderfully fitting irony that the Professorial Chair in Eugenics he founded (with his own fortune) at University College, London, is now a Chair in Genetics occupied by that wonderful man, the geneticist Steve Jones, an anti-hereditarian snail specialist.

In Galton's eyes, Ancient Greece and particularly the city state of Athens made for a cautionary tale. The Athenians symbolized a Golden Age of human culture and Galton believed that they lived such productive and admirable lives because of an unconscious form of natural selection. The city attracted the best and brightest from foreign lands, injecting fine genetic stock that bolstered the abilities of the existing population. More importantly, there was little dilution of the bloodlines of the upper and middle classes from the 'lower orders', the reason being slavery – very few slaves were known to have children with their masters – and so their genetic isolation protected the racial purity of the 'high Athenian breed'.

But if that was the case, why had the Golden Age ever come to

* The road to Hell truly is paved with good intentions. It is worth remembering that many Nazi doctors had 'altruistic' motives. They thought they were doing something good – benefiting society – forgetting in the process the autonomy of the individuals they were intent on experimenting upon.

an end? Galton suggested that the reason was that marriage eventually became less fashionable and fewer of the Athenian elite had large families. The population was maintained by accepting more immigrants, though they were increasingly of a poorer genetic quality. Athens sank into moral and economic decline.

Galton drew important lessons from the rise and fall of Athenian culture. He saw the damage that could be done to a society by paying no heed to the breeding practices of the population. He staunchly believed that the great human qualities were precious commodities and should be nurtured in the same way as cattle are bred for their health, the quality of the meat and their resistance to disease.

So it was that Darwin's considered and closely argued theory was kidnapped by Spencer, Galton and other theorists of the time who became known as the 'Social Darwinists'. They found that natural selection could be turned around in support of their own extreme social prejudices. Racism, intellectual snobbery and misguided nationalism, none of which was an unusual quality in a Victorian intellectual, lay at the heart of the new science of eugenics. From our privileged twenty-first-century vantage point, we can see that their analysis of human civilization was jumbled, pseudo-scientific, and often just plain wrong; it would end up being used to support the sterilization and degradation of the physically handicapped or mentally ill, and the 'removal' of genetically 'inferior' members of society. The rest is history – a history most of us would rather, but must not, forget. By the 1890s, critics were calling Spencer an 'extinct megalosaur' and his work was largely discredited, but the road first travelled by the original Social Darwinists led all the way to the enforced sterilization of the mentally ill in California, and eventually to the watershed of modern morality: the gates of Auschwitz and the pursuit of the Nazi Aryan dream. The Social Darwinists' theories were, in hindsight, equally as foolish and almost as threatening as Hitler's own ideology.

The truth is that Spencer, Galton, Huxley and the other proponents of such theories made several fundamental mistakes. The first was that Darwin's concept of 'fitness' does not match anyone's idea of admirable human 'qualities'. There is no reason why in modern society moral rectitude, intellectual vigour and physical capability should translate to increased reproductive potential. More to the point, we now know that contemporary human society most certainly does not operate according to the laws of natural selection. Among *Homo sapiens*, natural selection depends on the transmission of human characteristics from one generation to the next, and the possibility that these traits, whether physical or mental, are slightly changed in the process. We now know that the mechanism by which this happens involves our genes, and in particular the mutation of an individual's gene which can affect that person's relative fitness in any given environment.

For natural selection to work, those with physical or mental traits more suited to their environment or lifestyle must have more children than those who do not. There are special cases whereby severe mental or physical deformities do prevent people from having children. For example, if a genetic mutation causes someone to be born sterile, then obviously that mutation will not be passed on. But if substantial genetic differences do exist in the normal range of traits such as intelligence, physical prowess and personality type – and for many of these it is not as yet clear what exactly the genetic components are* – then in modern societies these differences do not markedly affect our reproductive potential. Intelligent, strong-willed or talented people do not on the whole leave more offspring in the world than the average; nor are accomplished athletes necessarily more successful at having babies than couch potatoes.

The Social Darwinists were also wrong to suggest that natural selection should act as a guide to the way we ought to live. This is an example of the so-called 'naturalistic fallacy'. There is no logical reason why there is anything morally worthwhile about the process of natural selection; it is, by its very nature, morally neutral, a biological process based on random mutations.

Now it's pretty obvious that 'survival of the fittest' doesn't describe the way we live in modern society, but for most of the evolutionary history of *Homo sapiens* things were very different. Survival of the fittest, the perpetual lottery of human survival, of

* Most of these complex traits are likely to be polygenic – that is to say, produced by the interaction of many different genes.

The Arecibo radio telescope is constantly being tuned and directed in the search for extra-terrestrial life. Do not humans have the greatest instinct for inquisitiveness of all animals?

Dr Seth Shostak/Science Photo Library

Opposite top: Whichever horizon you look towards on this part of the Kenyan Rift Valley, the cradle of human beginnings, there is now no obvious sign of human life. Yet seconds after this photograph was taken, the mobile phone in my left trouser pocket rang – Richard Dawkins's office from Oxford was on the line.

As many paintings and suits of armour of the period show, size certainly counted in the sixteenth century. Was this outfit worn by the Duke of Urbino (1503–72), painted by Agnolo Bronzino, intended to attract the ladies or to intimidate his male competitors?

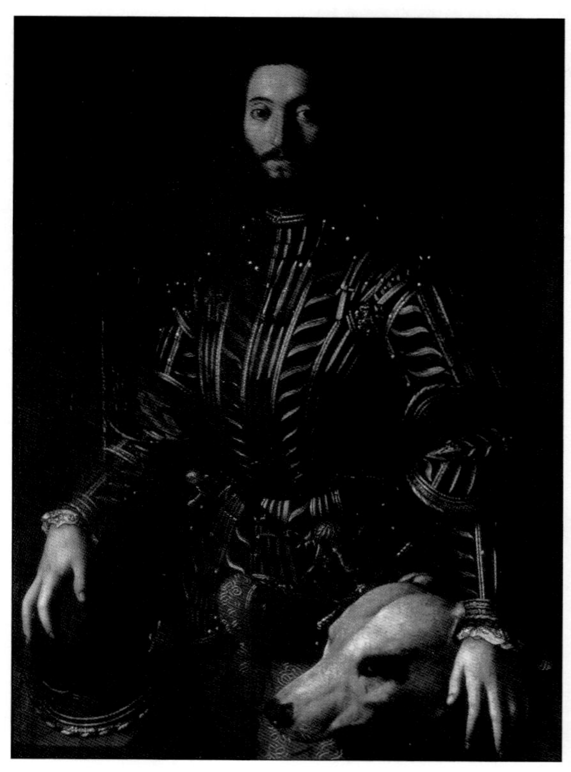

Opposite bottom: The toughness of life on the African savannah meant that calorie-rich, high-fat food was at a premium. It has left a craving for some unhealthy foods.

I spent many sleepless nights wondering about this drake. Faced with such a well-endowed partner, wouldn't be surprising if the female complained of occasional headache (see pages 186-7). Kevin G. McCracken

Left: Each of the T-shirts inside thes jars was worn for two nights by a female undergraduate and then smelt blind by me, after which I ranked them in order of appeal. Craig Roberts of Newcastle University knew the girls' tissue typ and scored these in order of their compatibility to my own. A score c 6/6 signified the closest match and 0/6 no tissue compatibility at all.

A legacy of the Battle of the Somme in 1916, there are thousands of graves like these near Arras, filled with bodies so mutilated that the dead cannot be identified. What instinct is it that leads men to senseless violence of this kind with the protagonists on both sides sure that they are in the right?

The record of my testosterone level – taken every ten minutes during the England–Argentina World Cup match in June 2002 – shows that it doubled immediately when Beckham scored his famous penalty, and rose again after victory was secured. I bet the England team had even higher levels by the end of the match. Reuters

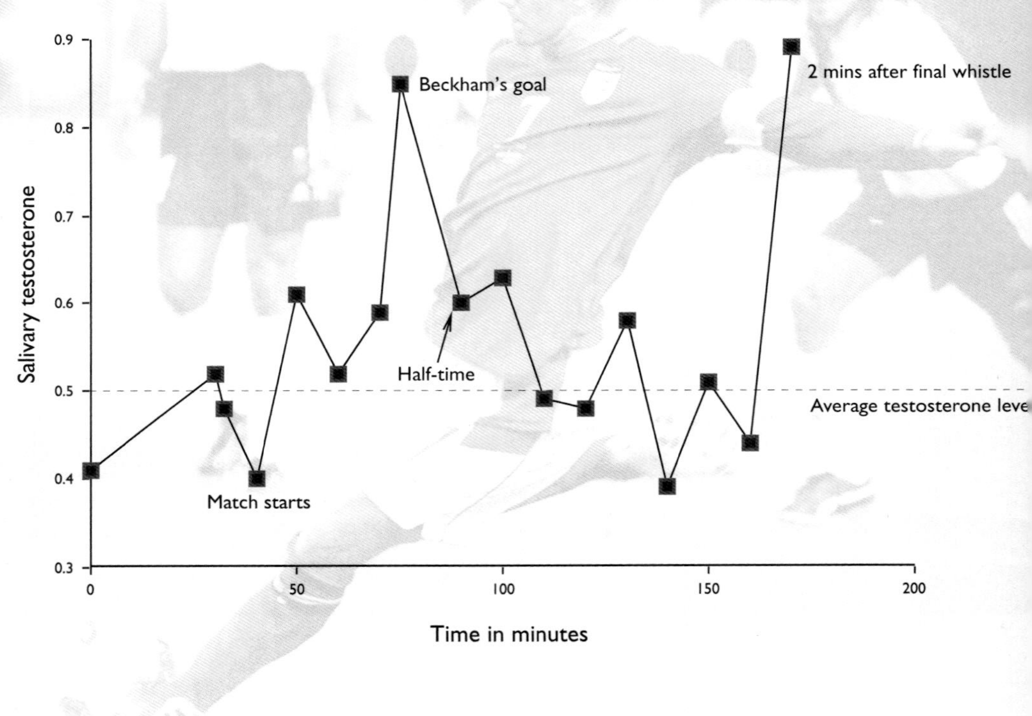

Opposite top: I am certain that I have many genes coding for competitiveness. I'm demonstrating this instinct by fencing with a member of the British Olympic team. Draw your own conclusion about which one is me.

Opposite bottom: I also have a dominant gene for cowardice, and gave up climbing for this reason some years ago. When doing this climb for the television camera, I spent moments frozen in sheer terror.

This cave painting, from a remote part of the Kimberley, North-Western Australia, is thought to be 5,000 – 10,000 years old. The prone figures of the dead woman and her infant – presumably signifying death in childbirth – testify that, like us, these people also felt sadness, anger and love, and that their spirituality was deeply ingrained.

human being pitted against human being, is an accurate picture of natural selection. Let us put the social theorists' moral ideals aside and examine the real factor that has historically driven evolution. It is competition, and it can be viewed from two quite different perspectives.

Firstly, there is competition between members of different species. If two kinds of hyenas – say Tough Hyena and Feeble Hyena – are vying to prey on the same antelope herds, drink at the same water hole and fight for the same territory, and one goes on to win out over the other, an entire species – Feeble Hyena, in this case – is wiped out. We know that many Feeble Hyenas have lived and died in evolutionary history. Still, a simple shift in environmental conditions – say, the drying up of a river, or an infestation by Hyena-biting termites, or the spread of a disease in gazelles or other prey – can trigger an either/or situation, rather like the balloon game in which one member must jump out to save the remaining passengers.

From this standpoint, it may appear as though species is pitted against species, a battle in which natural selection will determine the victor, but this is a view like that of a sonar image which can pick up the outline of an enormous fish swimming through the ocean but is unable to discern the individual members of that shoal. The fact is that the engine that drives evolutionary change is not competition between species. Remember that evolution can't act on a whole group; only individuals can be 'fitter' and more successful in surviving and reproducing, and thus do better at passing on their (mutated) genes to the next generation. For any given evolutionary change, there has to be competition between one individual Tough Hyena and another individual Tough Hyena. That's exactly how they got to be Tough in the first place.

Competition between members of the same species is the engine that drives natural selection. Think back to the explosion in brain development which separated the *Homo* lineage from that of the great apes. Each step along the way – and they could have been tiny steps or larger steps; we're still unclear on the details – was a product of competition between contemporaries, humans who breathed, ate, made tools, perhaps even talked to one another. They were competing for mating opportunities and for resources such as food, water or shelter; for the chance to survive and further their genes by reproducing. Can we see the imprint of this constant conflict, rivalry and competitive game-playing in ourselves today? Over aeons of evolutionary time, it has indeed left its mark on our psychological make-up, and we can see the results in all areas of human life, from the physical tussles of !Kung adolescents to the cut and thrust of the financial markets. In modern society, survival of the fittest does not mean that *only* the 'fittest' survive, but it does mean that we sometimes act as if that were the case.

'The meek shall inherit the earth,' said J. Paul Getty, 'but not the mineral rights.'

Your first ever conflict

The vanguard of the 'New Darwinists' in the 1960s and 1970s might not have been rewarded with material wealth, but they did have the satisfaction of transforming the world's view of evolution. William Hamilton, as we know, carved his brilliant theory of kin selection and inclusive fitness into the hard stone of mathematics. Two years later, George C. Williams published his classic book *Adaptation and Natural Selection*, which turned Hamilton's equations into prose, allowing biologists everywhere to understand its implications and continue the challenge against the dogma of group selection.

Williams's book was a bible for the third giant of the evolutionary revolution, Robert Trivers, while he was a graduate student at Harvard. Trivers is best known for his work on altruism, a subject we will examine in later chapters, but in 1974 he published a seminal paper that made a surprising and controversial prediction. It was based upon the theory of kin selection and demonstrated that conflict between human beings begins before we are even born.

We tend to imagine the co-existence of a mother and her unborn baby as a model of co-operation and harmony. The fetus grows happily in its bath of amniotic fluid, coddled at a perfect temperature, protected from shock and trauma by a layer of fatty tissue that surrounds the expanding uterus. Snaking its way into the heart of this cosy nest is the umbilical cord, providing a nutrient- and oxygen-rich supply of blood to the developing baby. But this fanciful image of the mother–child union hides a more complex reality. The mathematics of kin selection predict that during the nine months of pregnancy the mother's body will inevitably be in conflict with the fetus, and vice versa. After all, the mother and baby are not genetically identical; in fact, they are related by only half the same genes. Moreover, because the baby carries genes from its father, it is not even immunologically compatible with its mother. Strikingly, it is a piece of foreign tissue. Implanted anywhere else in the body and at any other time, it would be immediately rejected by the mother's immune system. The baby *in utero* is almost the perfect parasite, and to this day the precise reason why it is not shed, like a transplanted foreign kidney, is not fully understood.

Leaving this strange and poorly understood relationship with the mother's immune system aside, there are evolutionary changes that help the fetus develop more effectively. But these are at the expense of the mother, and therefore at the expense of brothers and sisters who may be born in the future. One example of this is the mother's blood supply through the placenta. When the embryo implants in the mother's uterus, it actually 'invades' the uterine wall, especially the walls of the maternal arteries, altering their structure so that they are no longer receptive to their usual control from the mother's body. These arteries can then carry their precious cargo of oxygen, hormones and nutrients directly to the fetus without the mother's body interfering. If the mother's body tries to restrict the amount of nutrients reaching the fetus, this will lower the levels reaching her own tissues. And because the fetus has plugged directly into the mother's blood supply, it is also able to secrete chemicals that are easily transferred through the placenta and around the mother's body; among these are hormones which act as the fetus's means of maintaining the pregnancy. They will compete with the mother's body for nutrients after she eats each meal and are able to trigger the placenta to produce enough progesterone to maintain the pregnancy even if one of the mother's ovaries shuts down completely. Meanwhile, the mother's body is frantically releasing its own growth factors to limit the fetus's invasion into her tissues.

So the effects of genes expressed in the fetus are countered by changes in the mother's gene expression. It makes perfect evolutionary sense. If the mother is in danger, so are all of her future children, and the danger has to be weighed against the well-being of the current inhabitant of her womb (see box). Two opposing genetic forces are engaged in an evolutionary battle. Biologists used to think that what's good for the mother is good for the baby. We now know that's not necessarily the case.

For those readers who are not shy of mathematics, the following is a brief example of the mathematics of kin selection in the womb.

Say a mother is pregnant with her first child. She is young and therefore has the potential to bear more children in the future. We know that a mother shares 0.5 of her genes with all her children (and this includes the children that will be born in the future). Similarly, a fetus shares 0.5 of its genes with all future siblings (assuming they have the same father).

Imagine that the mother carries a genetic mutation that benefits the survival of the existing fetus at the expense of her own health and survival chances. Not just the mother will suffer from this mutation – she's likely to have fewer healthy children in the future. This effect on future offspring is effectively a debit from her genetic account. It is as though the mother has made the following calculation.

Assume that the mutation increases the chance of the present fetus surviving from 80% to 90%, but decreases the chance of future babies surviving from 80% to 75%. Assume, in this hypothetical example, that the mother could have three more babies in her lifetime. Therefore, the overall genetic benefit for the mother will be: the proportion of genes in common with the present fetus multiplied by the percentage benefit to the fetus, or 0.5 x 10%.

The genetic cost to the mother will be the number of future babies multiplied by the proportion of genes in common with them multiplied by the percentage cost to future babies, or 3 x 0.5 x 5%.

Therefore, the net benefit is: (0.5 x 10%) – (3 x 0.5 x 5%) = –2.5%.

Because the outcome is negative, the end result is a cost rather than a benefit – for the mother. In the long run the mutation, if it occurs in the mother, will not be selected for. But what if a mutation occurs in the fetus which has the same effect on its survival?

The fetus shares 100%, or 1.0, of its own genes. From the fetus's point of view, its own survival is twice as important as its siblings' survival. So the new equation would be: (1.0 x 10%) – (3 x 0.5% x 5%) = +2.5%.

The outcome is positive, so there is an overall benefit to the fetus and the genetic change will in the long run be selected. Because one result is positive and the other negative, there will be a constant tension between the two opposing forces. In this case, the net result would be zero.

A genetic tug-of-war

We've seen that conflict is tightly bound up in the most fundamental of human processes, even the development of a fertilized ovum into an infant. Over the long term, the fight over the blood supply and the nutrients it contains is literally a life-or-death struggle, but within the developing fetus another powerful skirmish is taking place. This time, the two combatants are the mother and father, and again the battle is waged using genes as their weapons.

Each cell in a normal developing embryo contains forty-six chromosomes in the form of twenty-three pairs. One half of each pair is from the mother, and half is from the father. Apart from the XY chromosomes which determine the sex of the child, there are no major differences in the function and order of the two sets of genetic code. We effectively have two copies of the instruction manual, and for each step of the development process we need to refer to only one of the manuals at any one time. Which manual is used is generally thought to depend on which of the two particular variants of each gene (called 'alleles') we have received is dominant. If one is dominant and its opposite number is recessive, then the dominant allele will win out.

It would appear, then, as if a complete half of our full complement of genes is effectively passed over. But as we know, genes have evolved to maximize their chances of survival, and the maternal and paternal genes have slightly conflicting interests during the gestation of a fetus. From the point of view of the father's genes, there's a strong interest in building a 'bigger' baby. Larger babies will be more able to survive the pregnancy, and when they're born they are more likely to be healthy and so will have a better chance of surviving infancy. But from the mother's point of view – or, strictly speaking, from the mother's genes' point of view – the baby should be healthy but not too large. An oversized baby will drain the mother of precious nutrients during the pregnancy and is much more likely to present a threat to her health when it's born. The chances of a mother dying during childbirth are actually quite high, and the bigger the baby, the higher the chances. So the mother's genes are looking for a happy medium between the size and the viability of the fetus.

Researchers at Cambridge University were amazed to find that these competing interests are occasionally at loggerheads in the developing fetus. The reason is that some maternal or paternal genes have evolved an extraordinary ability to turn themselves 'on' during the development of an embryo, whether or not they are dominant or recessive. The process is called imprinting. A very small number of genes appear to be 'marked', or imprinted, according to whether they came from the mother or father, and this marking triggers the gene to become active at a certain point in development, whether or not it is dominant or recessive. It's a little like highlighting passages in just one of the two instruction manuals in order to catch the reader's eye. The surprise is that the maternal and paternal forms of the same gene are not necessarily pulling in the same direction, proving that our evolutionary suspicions were right. There is indeed a genetic tug-of-war for survival going on between mother and father.

The Cambridge team discovered that one such imprinted gene, IGF2, makes a growth hormone, rather similar to insulin, which has a role to play in the development of the cells in the early embryo and the placenta. When the hormone becomes active in the placenta it increases the amount of nutrients that can be passed through from mother to child. It is not a surprise, therefore, that this gene is paternal.

All mammalian genes share this technique of imprinting, so the researchers tested their theory on mice. They 'knocked out' IGF2 in a developing mouse embryo, effectively rendering the gene completely ineffective. Sure enough, when the mouse pups were born they were a remarkable 40 per cent smaller than normal.

It seems the father is delivering an imprinted gene that strives to increase the growth rate of the developing fetus. Then the researchers realized that maternal genes seem to be working in the opposite direction. The mother switches on a gene for a receptor whose role is to mop up excess protein produced by the paternal gene IGF2. In effect, her genes are countering the effect of the paternal IGF2, reducing the amount of nutrients getting across the placenta. Those mice that lacked the maternal gene, and the receptor to stop the paternal IGF2 doing its work, were 16 per cent larger than normal.*

This process of imprinting reveals that parents are engaged in a battle for control over the development of their baby. The outcome, it should be stressed, is not a happy compromise which aims to reconcile these conflicting interests. Remember that natural selection never results in a compromise for more than one person; rather, it acts on the genes of one particular individual and rewards that individual's success. As we've seen, the outcome is the end-product of an evolutionary arms race, a struggle which has been waged between the survival chances of the present baby and the survival chances of future offspring (and therefore the survival chances of the mother).

Without the benefit of modern medical technology, childbirth leads to a high mortality rate for the mother, partly because the size of the average newborn does pose a grave risk. According to the World Health Organization, 1,600 women die around the world every day giving birth – that's over half a million every year. And in the developing world, where medical resources and Caesarean section are not always available, a good proportion of these deaths can be loosely associated with the relatively large size of the human baby and its consequent need to have a very good blood supply. It seems that the efforts of paternal genes at imprinting have played a role in these grim statistics.

* Work in my own laboratory shows that even in the human egg, three days after fertilization, this gene has started to work. This may be why cloning leads to the production of overweight animals which have severe growth disturbances, because after cloning, both chromosomes of a pair are inherited from one parent. The pattern on imprinting is thus impaired, with dire consequences for the baby. This is one good reason why most scientists are implacably opposed to human reproductive cloning.

The competing male

'No instinct has been produced for the exclusive good of other animals,' said Darwin, 'but each animal takes advantage of the instincts of others.' Darwin realized that instincts are not 'designed' to make us better people, nor are they designed to promote the good of the species as a whole. They come about only through the single-minded pursuit of individual genetic success, so any traits that enhance the chances of survival and reproduction of those genes are going to do well and proliferate.

Where competition for mates is concerned, nature has shown that we are likely to develop a tendency for competitiveness and aggression. These traits are a product of 'sexual selection'. Natural selection, it is worth reiterating, is the process whereby traits are selected which enhance the survival of any particular individual; they may include spatial awareness, which helps our ability to track down and kill prey, manual dexterity, useful for tool-making, fear of heights, and so on. Anything which allows us to live longer, eat better and protect ourselves and our kin from danger so that we and they can reproduce could be a possible adaptive mechanism and a product of natural selection. It's what allowed us to become adapted to our environment, and in our case, that was the savannah.

Sexual selection, on the other hand, is the process that favours your chances of survival if you have an increased number of mates, and therefore a larger number of offspring. Anything that increases your chances of having sex will be adaptive by sexual selection. Many species exist in which males directly compete for mates and they develop weapons whose sole purpose is to engage other males in battle. The antlers of red deer are a good example, because we know that they're used to fight over females rather than to fight predators. Darwin said, 'The development of certain structures - of the horns, for instance, in certain stags – has been carried to a wonderful extreme; and in some cases to an extreme which, as far as the general conditions of life are concerned, must be slightly injurious to the male.' Bad for the species, because the more lethal the weapons, the more deer will be injured or killed. What's more, the energy and extra nutrients required to grow the antlers are substantial. The race to grow bigger antlers will do the species as a whole no good whatsoever in adapting to its environment, escaping from deer-hunting leopards, or just possibly tracking down new sources of food. But large antlers are just right for the individual stag that wins the fights. At its heart, sexual selection is all about competition and conflict within the species.

For animals that have adopted the harem system, male conflict is an ordinary part of social life. Harem rights, and the vast genetic rewards that go along with these rights, will be determined by a fight between one or more males. Elephant seals are so large because of their fighting prowess; the biggest, fattest seal wins the fights and becomes the oversized master of the harem.

But sexual selection is not all about clashing antlers and Sumo-like shoving matches. Darwin wrote in *The Descent of Man*, 'the power to charm the female has sometimes been more important than the power to conquer other males in battle'. 'Charm' takes a more subtle and roundabout route than brute force, and as we humans know all too well, it can have the oddest consequences.

The peacock's tail

Where charm is concerned, sexual selection is powered by the opposite sex. In most species that practise sexual reproduction, the females are in a position to choose their mates. The fact that some females may be unavailable as a result of pregnancy or taking care of their young means that there will generally be an imbalance of males to females. And let's not forget that the females are going to invest precious eggs and also far more time and energy in bearing and bringing up their offspring than the males. So there's no room for making mistakes when it comes to picking your man. In other words, the choice females make is a powerful one; they're not about to sit on the sidelines waiting for the menfolk to fight it out.

Females prefer certain physical traits in a male, and the more pronounced that trait, the more attractive the male will appear. This explanation is intentionally circular; the preference for bright colours, tuneful songs or displays of dancing or fearlessness need not be based on any *practical* value. And as any woman intuitively knows, despite the sexy fast cars, cocktails and acts of heroism, James Bond probably never phoned when he said he would.

In nature, the classic example is the peacock's tail. The glorious, iridescent colours and oversized sweep of the peacock's train are an extreme manifestation of sexual selection. Peacocks arrogantly strut, fanning their plumage, their feathered eyes winking and shimmering in the light. It's no surprise that the word 'peacock' is rightly a symbol of the dandy and male narcissism.

The tail may be attractive, but it's an enormous burden. It has no practical use. When the peacock is under attack, or fighting with other peacocks, the tail is tucked away out of danger. Owning such a capricious adornment is rather like wearing a white tie, starched waistcoat and tails in a bar brawl; dressing like Fred Astaire in a suit that is constricting, pointlessly elaborate and bound to get ruined simply shows from what origins male fashion has evolved. The peacock's tail, too, is a significant handicap and an obvious disadvantage when it comes to day-to-day survival. It hampers speed and ability. It requires large amounts of energy to grow and maintain. But the biggest drawback of all is the tail's visibility to predators. When it's unfolded, the peacock may as well be wearing a flashing neon sign that says EAT ME. Even a pretty average evolutionary accountant would have advised the peacocks not to bother; indeed, *natural* selection would never have favoured them.

Sexual selection had other ideas. As soon as one peacock happened to grow a slightly larger or more colourful tail, and as soon as it caught the eye of an admiring peahen, sexual selection saw to it that the rest of the species had to follow suit. In a competition for mates, there can be no holding back. During the evolution of the peacock's tail, natural selection and sexual selection were, as it were, in constant tension, and sexual selection won. The female preference for larger tails meant the 'large tail' gene was bound to do well. Similarly, once one male deer inherited a chance mutation gene that produced larger antlers, there was a good chance that the gene would spread through the population. (It is certainly possible that in many other species that do not have these mate-attracting encumbrances, similar attributes either died out or did not get off the ground.)

Yet clearly the paradox remains. Why do females prefer a trait that has no apparent survival advantage for their offspring? The great 1930s biologist Sir Ronald Fisher described the process as 'runaway sexual selection' – a vicious circle that ends up building magnificent mate-attracting traits which may, at the same time, be a handicap to survival.

For Fisher's theory to work, there has to be a starting point, a moment in evolutionary time when peahens began to prefer longer tails. This preference could be completely random, the result of a single arbitrary preference – say, one female with a kink in her sexual psychology. The long tail may signify nothing. But what if the male with the long tail, by chance, happens to have particularly good genes in other departments? Maybe he is swifter, bigger, or has a better immune system. In that case, the preference for large tails could be a successful tactic, because he's likely to father healthier and fitter offspring. In the short term the gene would spread throughout the gene pool, because longer tails started off as equivalent to better genes, even though this link may only be temporary. If enough females inherit the preference for males with longer tails, then their female offspring will also inherit that preference, and of course their male offspring will have a longer than average tail. This hypothesis has become known as the 'sexy sons' theory; male offspring with long tails are sexier to the next generation simply because the mate preference traits are inheritable.

Since Fisher outlined his ideas some seventy years ago, biologists have computer-modelled his scenario and found it to remain intact over many generations. Their models are simplified, inevitably, but the principle seems to be sound; sexual selection can indeed incrementally overcome natural selection. The peacock's tail, a product of a vicious circle, means every male has to keep up, or lose out.

There are some rather unexpected results of runaway selection. Most male birds, for example, do not have a penis. They mate via a 'cloacal kiss', which involves the male and female genital openings touching briefly to transfer the sperm. Some lucky species of ducks and ostriches do have penises, however. Recently, scientists at the University of Alaska reported in the most authoritative journal, *Nature*, a specimen of the Argentine Lake Duck with a penis nearly half a metre long, the same length as its body. Moreover, it is shaped like an overlong corkscrew. The scientists suggest that this memorable appendage is the result of Fisher's model of sexual selection. Moreover, this species is known to be especially promiscuous and boisterous in its pursuit of sexual opportunity, which means that runaway selection, through an arbitrary female preference for a longer penis, could have even more opportunity to exert pressure.*

Good genes

We find the pointlessness of the runaway selection theory intuitively disconcerting. Evolution was supposed to solve problems, not create them. Perhaps this is the (unconscious) motivation of biologists who are convinced that a long peacock tail was originally, and still is, an indicator of good genes. They believe that an animal could not be so decadent and wasteful as to grow an enormous tail if it did not provide evidence of some other, more worthwhile aspect of the male genome, like strength or fertility.

To be fair, they do have some evidence to back up this claim. In one or two cases male flamboyance does indeed appear to go hand in hand with a real genetic advantage, beyond the fact that their male offspring will be more attractive to the next generation of females. Two species of bizarre-looking flies in the Diopsidae family, found in the African tropics, have eyes on the end of comically long stalks sticking out at right angles to their head. These stalks can extend to one and a half times the insect's body length. Every evening the males stake out rootlets along a stream bank and every so often they will face off with another male, eye to eye. The fly with the shorter eye stalk will usually back off, just as small-antlered deer will usually back down from a confrontation. The females, who meander through the morass of skirmishing males, will pick the choicest specimens – the ones with the longest eye-stalks – and mate with them.

* The first author on the paper, Dr McCracken, acknowledges that many questions remain. How much of the penis does the drake actually insert? And does the anatomy of the female make them unusually difficult to inseminate?

It is true that her offspring will have longer eye-stalks and therefore be more successful in dominating the Diopsidae singles circuit for the following generation, but there is a more important benefit. Among these species of Diopsidae, the sex ratio is biased towards females, by a ratio of two to one. The long-eye-stalk male flies have an especially potent Y chromosome that, unusually, has the effect of producing more male offspring than female. For his female partner, having more sons in the next generation is an advantage; in a population dominated by females, it maximizes her genetic output. Because this potent Y chromosome appears only in a proportion of the males, the female choice is a very real choice between inferior and superior genes.

Competition between males can be more overtly linked to their genetic fitness. Some species of birds, for example sage grouse, have developed a springtime ritual called 'lekking'. A lek is a communal mating arena in which all the males will gather to strut, posture and dance for the benefit of the females. The females will wander through the gyrating crowd and have been noted to observe all the males' displays, apparently comparing them before they make their choice of mate. Like John Travolta in *Saturday Night Fever*, a great performance will virtually guarantee a mate at the end of the night – several, in fact, for the best dancer. In this case, that's because the display itself is proof of a genetic advantage. Those males with the best physical display are likely to be stronger and healthier than the poor lonely souls who shuffle gloomily around the fringes of the lek.

And there's another interesting process at work: among sage grouse, just as with humans, some guys get all the girls. The best grouse dancer can end up with as many as 80 per cent of the females choosing him. Recent research has shown that that may be because some of the females are relying on the other females' preferences to lead them to the best male. Using stuffed dummies to represent interested grouse females, the researchers have shown that a female grouse tends to choose a male who appears to have other females in his territory rather than one of the less popular males. Likewise, in a human psychological study, women appeared to be more likely to express an interest in going out with a man if they were told that other women also find him attractive. Perhaps copying others in choosing the mate with the best genes saves time and energy, and provides a little reassurance.

All this does little to convince the followers of Fisher, the believers in runaway sexual selection, who are perfecting their mathematical models and demonstrating the essential pointlessness of the peacock's tail. Nonetheless, the opposing faction is sticking by its guns and occasionally finding species which adhere to the Good Genes theory. It seems increasingly likely that both theories have something to recommend them. The Good Genes theory provides a reason why females first develop a particular preference, whether for a mating dance, a certain style of birdsong or a brightly coloured tail; the Runaway Selection theory explains how this preference leads to a trait that can get out of control, eventually straying far beyond any sensible genetic use.

The human tale

Human beings have not grown eye-stalks, nor do we possess an ability to unfurl some elegant tapestry of feathers, nor even a remarkable corkscrew, but we do possess one physical attribute that is more extraordinary than even the glossiest and most colourful tail fan: the human brain. We've already seen how difficult it is to account for the phenomenal growth of the human brain. The fossil record does not show a parallel boost in human ability or creativity; the stone flake and then the teardrop-shaped handaxe remained unchanged for around a million years each. During this time, the brain grew, and grew. Equally, there is no obvious environmental or climatic shift that would have provided the impetus for such an adaptation.

Fisher might unknowingly have given us the answer. Perhaps the brain is a product of runaway selection; in other words, brains made us sexy. Those early humans who were more adept at communicating – whether through grunts, expressions, or a fully formed language – would be well placed to find a mate. The exchange of information, of gossip, and even joke-telling might have been a feature of life as far back as *Homo erectus*, one and a half million years ago. That social intelligence played a role in increasing brain size is an attractive theory, and it makes intuitive sense. Any advantage we could get from manipulating others and creating alliances would prove useful to our chances of having sex. Our primate cousins prove that dominance hierarchies can make or break an individual's success in passing on his genes. No-one messes with the alpha male.

Others argue that bigger brains were needed for survival. For example, we might have used tools to find food in previously inaccessible places, just as chimpanzees have been known to 'fish' for termites in termite mounds using long blades of grass. But we have to bear in mind that most other animals have survived and prospered without the need for brains that are wildly out of proportion to the size of their bodies. Brains are only as big as they need to be; for some reason ours hitched a ride on an evolutionary skyrocket. Brains are so expensive to build and maintain that the selection pressure must have been great. Like the peacock's tail, the brain might well have proved a liability.

But there's one point about the brain which runaway sexual selection cannot explain. By the very rules of sexual selection, males should be the ones who develop highly complex ornaments with which to woo females, while the females themselves devote their energies to choosing a mate and have few ornaments of their own because they don't need them. But we also know that modern men's and women's brains are extremely similar in size and capacity; they're only very slightly proportionately larger in men owing to size differences. If men developed significantly bigger brains to charm and woo women, how come women's brains also evolved to be the same size? It seems our oversized brains can't just be ornaments of sexual selection. There must be another explanation.

Machiavellian instincts

A five-hundred-year-old book from the great northern Italian city-state Florence has proved remarkably influential on modern theories of human evolution. Niccolò Machiavelli's best known work, *The Prince*, which was written in 1513 but published posthumously in

1532. Its theme is essentially that princes should retain total control of their territories using any expedient means, including deceit. Machiavelli was an arch advocate of the adage that the ends justified the means. In several sections of *The Prince* Machiavelli praises Cesare Borgia, the notorious and much despised tyrant of the Romagna region of northern Italy. The book is a primer on manipulation and control, giving its readers a set of rules on how to employ under-hand tactics, to lie and cheat in order to get what we want. It tells us we need to play a role – for example the 'benevolent leader', in the case of the Prince himself – that allows us to throw a veil over our darker motivations and rampant self-interest: 'It is useful, for example, to appear merciful, trustworthy, humane, blameless, religious – and to be so – yet to be in such measure prepared in mind that if you need to be not so, you can and do change to the contrary.' Now, two biologists at St Andrews University in Scotland, Andrew Whiten and Richard Byrne, have applied Machiavelli's musings to human interaction on the savannah.

Their theory is based on the 'social intelligence hypothesis', which asserts that dealing with human relationships demands greater intelligence than tool use or the day-to-day trials of survival. For example, our forebears, who lived their entire lives in a small, tightly knit group, needed to read the intentions or desires of others and convince them of certain courses of action. They had to understand the social power structure of the group and of kinship. Making alliances, a process stripped bare in reality TV shows such as *Survivor* or *Big Brother* as well as in accounts of the behaviour of other primates, would also be crucial to one's success. You can see how devious individuals would be much better at securing more than their fair share of resources, assuming they did not get caught. In a small group, getting caught lying or cheating is extremely damaging to one's future prospects.

More brainpower is required to be a manipulator than to be manipulated. Whiten and Byrne have concluded that Machiavellian behaviour was the impetus for the upward evolutionary spiral of the human brain. Once a form of behaviour appears that exploits the gullible, there will be a selection pressure to match this exploitation with counter-exploitation, something that requires greater intelligence. The arms race would continue, turning brainpower into an essential evolutionary property, each Machiavelli being outdone by the next. Intensive work with primates and other animals has shown Whiten and Byrne that considerable brain capacity, particularly enlargement of the cortex, seems to be necessary for the efficient memory individuals need in order to build extensive social knowledge and rapidly learn new tactics in interaction.

If these tactics were indeed such a fundamental aspect of human evolution, our social world appears to be founded on exploitation and enmity. It is certainly the case that the greatest of friendships can be tainted by exploitation or rivalry. Banquo thought his 'noble partner' Macbeth was like a brother, 'bosom franchised and allegiance clear'; little did he think that he would end up in a ditch with twenty gashes in his head – though his instincts got him worried near the end.

One may also extend the theory to sexual conquests. How much of seduction is Machiavellian? To what degree is sex secured by misinformation and manipulation? We think sexual politics and social mores are complex enough in the modern world; half a million years ago perhaps they were just as difficult. Even if we want to put the darker side of human tactics aside, perhaps sexual selection does have an answer for why both men and women have equally sized brains. The human brain itself can be seen as the mechanism of sexual choice; perhaps as men evolved more highly complex brains to woo females, women too had to develop bigger brains in order to understand and choose their man. It seems that a complex combination of sexual selection and the need to interact with others socially and to make use of a certain Machiavellian intelligence may well account for how we've ended up with such large brains.

Human behaviour

Human beings lek. In our own way, we all try to dance like Travolta, shaking our tail feathers and clashing our eye-stalks, trying to win points on the sexual selection score board. But an impressive intellect or a fast wit doesn't necessarily cut it on the dance floor.

Consider the luck of Woody Allen, who asks a girl in the disco, 'What are you doing on Saturday night?' 'Committing suicide,' she says. (Woody replies, 'How about Friday?')

Let's briefly remind ourselves of the most desirable features in a mate. Women, on the whole, prefer high-status, ambitious males. How do these men show women their status? They display symbols of their financial and social superiority. I am reminded of my trip to Los Angeles to make the programme on sex in the *Human Instinct* series. The BBC had arranged for four fashionable women to sit at an outdoor café on a smart Beverly Hills boulevard. They were given cards to score the next male to drive up and sit at the next table on the sidewalk. Unknown to them, my producer had persuaded the café owner to keep the table for me. I drove up in an old banger, a Ford obtained from one of the local film studios. I wore a jumper and torn trousers, with dirty sneakers. When the ladies held up their cards like ice-rink judges, my four scores were 0, 0, 1, 0, average 0.25.

The next day I had had a manicure at a beauty salon and a shampoo and haircut done by Laurent, who does Gwyneth Paltrow's and Meg Ryan's hair. I wore a twenty-five-thousand-dollar set of cufflinks borrowed from Harry Winston's, a thirty-thousand-dollar diamond-studded watch from Cartier, and a Versace suit and tie from Rodeo Drive. And I was driving a brand-new open-top Bentley worth a quarter of a million dollars. As I walked past the same table and sat down, the cards went up: 10, 9, 10, 10, average 9.75. But when I got talking to one of the ladies afterwards, she said, 'But you know, really I liked you better yesterday. You weren't so Los Angeles.'

Women also look for heroes; the unselfish and the altruistic. Showing off these qualities is a permanent part of the male courtship display: the gentlemen is, traditionally, supposed to hold open the doors as well as reach for the bill when the time comes.

To put it crudely, the car, the expensive clothes and the understated confidence are indicators of a quality mate. Women do want the physical qualities too – the outward signs of strength, of virility and of health – but these are secondary considerations. For humans, then, the peacock's tail comes in the form of risk, prowess, status and wealth. It's all about consumption, showing off and confidence. Women want men who, as the Russians used to say, treat the sea as if it were knee deep.

Women, too, display symbols of their appeal as a potential mate. Dress or make-up often reflects biological traits men generally find attractive; for example, the hourglass shape of Victorian corsets mirrored the desire for the right waist-to-hip ratio, and rouge and lipstick mimics the increased blood flow of sexual desire. But what is interesting is the lengths women will go to in their dress and appearance even where there are no men present. Competition within the sex is just as important as their attractiveness to men.

Every member of every human culture would recognize a human instinct to show off and to waste resources on extravagance and frippery. All these qualities are a fundamental part of sexual selection, responsible for some of the more outlandish quirks of human behaviour.

The American and Canadian Indians of the west coast set the standard for excess. Potlatches were social occasions given by a host to establish or uphold his status position in society. Often they were held to mark a significant event in his family, such as the birth of a child, a daughter's first menses, or a son's marriage. This long-standing custom is a fine example of conspicuous consumption. The original purpose of these gatherings was probably to trade goods with neighbouring tribes, but as the wealth of the clans increased so did their tendency to give away valuables or destroy them. Great political gains could be made at these meetings, but at the expense of great wealth; it became accepted that only through huge extravagance could a clan win status and political power. Slaves were killed or set free. Hundreds of copper cooking pots were thrown into the sea. Enormous stacks of blankets were set on fire.

In modern society, with consumerism run rampant, extravagance requires creativity. Last year, realizing that mobile telephones were so commonplace that even the most expensive were no longer exclusive, the London jeweller De Grisogono began selling limited-edition phones encrusted with diamonds. They cost between fifteen and thirty thousand pounds. Apparently, signs of the zodiac are a favourite design.

Taking risks

At the time of writing, Larry Ellison is the second richest person in the world, next to Bill Gates (after the dip in Microsoft's fortunes during the US government anti-trust case against the company, Ellison briefly held the title of the world's richest person). Ellison, who made his fortune with the computer software company Oracle, is worth some fifty billion dollars, depending on the vagaries of the stock market. This personal fortune is worth more than the gross domestic product of Ecuador or Tunisia.

Ellison is described by his biographer as 'a one-man amusement park, Larryland'. He recently spent forty million dollars on a mansion in California, built in the style of a Japanese imperial palace. Bill Gates is known for humbly flying economy class; Ellison has bought his own SIAI-Marchetti fighter plane and indulges in mock dog-fights over the Pacific with his son (he tried to import a Russian MiG fighter jet, but the US government refused him permission, on the basis that it would be classified as a firearm). And while Bill Gates is known to enjoy the occasional day of recreational sailing with his family, in 1998 Larry Ellison won the Sydney-to-Hobart yacht race, beating eighty-mile-an-hour winds and twenty-foot waves that killed three fellow competitors.

Herodotus said that great deeds are usually wrought at great risk. I would add to that: great risks are often taken without any great deeds in mind.

But that is not to diminish the human achievements that owe their existence to a willingness to take a chance. Medieval Venetians bravely yet eagerly ventured into undiscovered oceans, running the risk of falling off the edge of a world they might have believed was flat.* Our will to survive is constantly undermined by the human ability to throw all caution to the winds. Nowadays, slot-machine addicts and high rollers alike are well aware that there is a vast

* I am not entirely convinced that medieval mariners truly believed in a flat earth. There is quite a lot of evidence which suggests that they considered the earth curved, at least. Still, long sea journeys then were unbelievably perilous and the odds on a successful return not great.

amount of money to be made out of this very instinct. Las Vegas is a neon advertisement for the fact that the house always has the advantage – and to prove it, casinos relieve Americans of sixty billion dollars a year. Many of their most loyal customers are poor and can scarcely afford to lose what may be next month's mortgage payment. Why do we gamble when we know the odds are against us?

Could the instinct to take risks be a perfectly useful adaptation? In this sense, it would have little to do with sexual selection or competition. Let's assume that the savannah is prone to occasional periods of scarcity. When the rainy season is cut short, for instance, fruit may be sparse and wandering herds of antelope rather thin on the ground. You know that the adults in the group are strong enough to survive; it has happened before, and hopefully soon the river will swell and animals will return to the valley. But there is a chance that your children will not survive. Perhaps you have young and vulnerable infants, or a baby whose mother may become too undernourished to breast-feed. The only chance of finding more food is to walk through a pass into the next valley. It is a long trip, with no guarantee of success. You may lose your way, starve, or be attacked by mountain lions. Perhaps you know of people in the past who made the same trip, and not all of whom returned.

Natural selection might well have favoured those who took risks like this one. It would be impossible to know the true odds of survival; we would not consciously weigh up the percentage chance of survival against the number of our children who could starve. Yet if, instinctively, the odds looked OK, many of us would indeed take the chance. But that does not explain Larry Ellison and his fighter jets. It does not explain why we would willingly swap financial security for 'just one more' one-in-a-million shot at a slot-machine jackpot.

The handicap of risks

Risky behaviour is not just a feature of human life. Consider the Trinidadian guppy. As we saw earlier, when a predator nears a school of guppies, one or two especially intrepid males will slowly approach the intruder, inspecting it for signs of danger. Generally they feel more disposed to do this when a female is near. Even so, why on earth would a defenceless guppy go nosing around a potentially life-threatening predator?

Last year, a German tourist was swimming off a beach in Thailand when he noticed a large tiger shark swimming close by. He furiously swam to the shore, grabbed his camera and waded back out again to take a photograph. The shark swam lazily over and sank his teeth into the man's leg. The bite severed an artery and the man bled to death. We can speculate as to whether or not he had intended to impress women sunbathing nearby, but this incident and the guppies's behaviour have more in common than you may think.

The two American scientists who spent many months watching guppies in the Paria River in Trinidad worked out that female guppies did indeed prefer to mate with those risk-taking males. In fact, laboratory studies have shown that when no females are around at all, no one male guppy will swim forward heroically to check out a potential predator more often than one of his male counterparts. Once again, sexual selection appears to promote behaviour that from the point of survival is at best futile and at worst injurious, even fatal. Natural and sexual selection appearing to contradict each other yet again.

Amotz Zahavi, Professor of Zoology at the Institute for Nature Conservation Research at Tel Aviv University and a legendary Israeli ethologist, has attempted to explain this behaviour. In the 1970s, he became entranced by the question of risky – and, from an evolutionary point of view, irrational – behaviour, and eventually presented a radical and clever theory which was immediately laughed out of court.

Costly behaviour is connected to the kind of 'signals' an animal is trying to transmit. According to the rules of sexual selection, a male guppy wants to send out a signal that says 'I've got great genes; you must try mating with me.' The problem is that even if the system begins with only those guppies with the best genes sending out those particular signals – say, displaying a certain colouration, or swimming with a certain lilt – then they're leaving themselves open to deceivers. Guppies with terrible genes could take advantage and sneakingly adopt the signals for themselves. Ultimately, this means that females would ignore the signals completely and use other criteria to choose their mate, or even choose them at random. The strategy is not stable; it cannot last over the long term and the signals become irrelevant.

Zahavi thought long and hard about this paradox. His chosen species of study for many years was the Arabian babbler, an undistinguished and unflamboyant bird living in the deserts of the Middle East. Just like the guppies, babblers take risks in the face of great danger. Through years of watching, waiting and thinking, Zahavi constructed an explanation for why some male babblers announce their presence to predators by shouting at them.

As we've seen, the peacock's tail is a 'handicap' – a burden and an energy-sapping accessory. Similarly, the risks undertaken by babblers or guppies may also be considered handicaps: approaching predators has an adverse effect on their genetic fitness because some of them are surely going to get caught out. Zahavi proposed that these animals are successful not just because females happen to have developed a liking for risk-taking babblers. That would make no sense; the negative effect on survival would counteract any benefit conferred by Fisher's runaway selection theory, because risky babblers can, and do, get eaten, if they are slow or unlucky. Instead, Zahavi suggested that the handicap existed because it was a true and honest indicator of the real genetic fitness of an animal.

The babbler can only afford to take the risk because he *has* better genes; he is healthier, fitter and faster, so his behaviour is a true measure of his desirability as a mate. Sometimes the risk does not pay off, even for the fastest babbler in the flock, but the strategy will be judged in the long term. The idea is to take the risk, but only when you know you can afford it, because then you know you are much less likely to get eaten. In other words, honesty is the best policy; not only that, but honesty is the only strategy which can last in the long term, because it prevents 'deceivers' rendering the strategy pointless. Those babblers with poor genes are more likely to get eaten if they take risks, so the 'deceiver' gene will not be successful in the long run.

Zahavi's theory has a name, the Handicap Principle, and it also applies to the intriguing behaviour of some mammals. Gazelles will sometimes stand up on their hind legs when a herd is stalked by a wild dog; in effect, that advertises their presence to the predator, a form of behaviour called 'stotting'. Why would a gazelle do this? Back when group selection was common currency, the answer seemed obvious: stotting distracts the wild dog and alerts the rest of the herd to its presence. It was explained as a classic act of self-sacrifice and a form of behaviour that could maximize the fitness of the group as a whole. No matter that one individual put itself willingly in danger, so long as the group benefited. But now we realize that for such a form of behaviour to evolve, it must help the individual, or rather the genes the individual carries. Enter the Handicap Principle. Stotting is energy-sapping, risky and, indeed, sometimes fatal. Precisely because of those factors, stotting provides reliable information about the quality of a mate, because only those gazelles who can afford to stot will do so. The handicap can only be borne by those with good genes, and studies of stotting gazelles show that they are indeed more likely to escape the wild dogs.

Initially, no-one believed this eccentric Israeli bird-watcher. Zahavi had devised a theory that, so it was thought, could be used to justify any kind of behaviour, no matter how deranged it appeared to be. The evolutionary biologist Robert Trivers teased Zahavi that his theory could even explain a species of bird which flew upside down to try to show to the opposite sex how successful they would be when flying the right way up. In an interview, Zahavi casually pointed out to the journalist Richard Conniff that some birds do in fact fly upside down as part of their mating display.

Zahavi never attempted to mathematize his theory, he expressed it in verbal form, just as I have done here, and often even more simply by saying things like 'something can be good, just because it's bad'. That would never have persuaded those ethologists and biologists who were beginning to rely heavily on game theory and formal proofs of the evolution of behavioural strategies. But eventually, Zahavi's theory was translated into a computer programme that could test whether or not the 'honest advertising' strategy could work in the long term, over many generations. It turned out that it could, and it did, but only when the handicap was costly. In other words, risks only pay off if the behaviour is particularly risky.

Living and writing at the end of the nineteenth century, the eccentric and irascible Thorstein Veblen was, it was frequently said, the 'last man to know everything'. He held a doctorate from Yale in moral philosophy based on a thesis about Immanuel Kant, and he spoke twenty-five languages. He was expelled from both Chicago and Stanford Universities, possibly, at least once, because of a sexual liaison of some kind, though of his lectures it was said 'He mumbled, he rambled, he digressed.' Though he was highly knowledgeable in literature, art, history, science, technology, agriculture, labour relations, teaching and industrial development, most people now classify Veblen as a political economist. He emphasized that humans, like Zahavi's birds, paraded their status. It was he who invented the memorable phrase 'conspicuous consumption'; it was directly linked in his mind to the notion of a person signalling their standing and reputation. 'Consumption of valuable goods', wrote Veblen in his book *Theory of the Leisure Class*, 'is a means of reputability to the gentleman of leisure.' He gave as an example 'intoxicating beverages and narcotics':

> If these articles of consumption are costly, they are felt to be noble and honorific. Therefore the base classes, primarily the women, practise an enforced continence with respect to these stimulants, except in countries where they are obtainable at a very low cost. From archaic times . . . it has been the prerequisite of the men of gentle birth and breeding to consume them.

In that light, a 'handicap' could be a tendency not just to make oneself a target to a predator, but to eat poisonous food, to exert energy on pointless and dangerous activities, or to spend precious resources without any discernible gain in survival or fitness. The blowfish or fugu is a highly sought after and expensive delicacy in Japan, but it can also be lethal. Its liver is deadly poisonous; it is the gourmet equivalent of Russian roulette. Chefs have to be highly trained and licensed to serve the fish, yet despite this precaution at least a hundred people die each year, most from ingesting unseen traces of liver tissue. Base jumping, a highly dangerous activity that involves flinging oneself off cliffs, bridges and towers and then opening a parachute, is increasingly popular. We drive faster than we need to, and more people than ever are now taking potentially dangerous recreational drugs.

Sometimes the link to sexual selection is overt. One experiment showed that men are more likely to take risks when crossing a road if a woman is watching. And people who live in tribal or hunter-gatherer societies are just as disposed to taking risks, like the Maasai of southern Kenya, whose young males are expected by ancient traditions to take part in a potentially lethal lion hunt before they can be eligible to marry.

But I suspect the instinct has now become so ingrained that risk is a fundamental element of human psychology, whether or not we are actively looking for a mate. Human beings delight in their excess, find more and more ingenious ways of putting themselves in jeopardy and adopt as many expensive handicaps as they can afford. We are living proof of Zahavi's theory.

The glass ceiling

Sexual selection acts more powerfully on males than it does on females simply because males have to compete more fiercely for mating rights. If the Handicap Principle is exhibited more strongly in male gazelles and babblers, surely we would expect to find it displayed more often in men than in women? From our everyday experience it does indeed seem to be the case that men do take more risks than women, both in their day-to-day lives and in terms of their liking for dangerous activities. Far more men than women drive too fast, for instance, and more men die as a result. This is one of the reasons why men have a lower life expectancy than women: they are involved in more fatal accidents of various kinds. On a more positive note, 90 per cent of the recipients of the Carnegie heroism awards are won by men. These awards go to heroic acts performed in the face of great risk.

Many psychologists and biologists would argue that it is nigh on impossible to *prove* that men are genetically predisposed to taking risks, rather than having picked up the habit from a persistent pattern of upbringing and sexual stereotyping. Just because a theory makes sense in the light of certain statistics - or makes a snug fit with another theory - does not necessarily mean that it's true. Others, though, are not so reticent in going out on a limb. Robert Wright, the author of *The Moral Animal*, believes that men not only have a greater willingness to take risks, they also have greater ambition, and that both are key evolutionary adaptations which are central to human 'maleness'.

If this is true, it leads to some interesting political implications. If male ambition is 'natural', could it be that differences in career success between the sexes are not necessarily caused only by sexual discrimination or prejudice? Broadly, according to this theory, women would not be expected to match men's success in any given field. Perhaps the fact that women's salaries are, on average, 75 per cent of the amount their male counterparts earn is a fair and inevitable result of their reduced ambition (we will, for the moment, put aside the effect of women taking time off to have and raise children). From this point of view, it seems, very controversially, that we shouldn't worry about inequalities in success or earning power.

But Wright suggests that his conclusions be used to support a more acceptable course of action. Few people would say that men are *better* at their chosen careers, simply that they tend to progress further and faster than women, thanks in part to increased ambition and a propensity to take risks. He says that because of this, women should be favoured by positive discrimination, not to counteract the effect of sexual discrimination, but to counteract the effect of human nature. The best possible outcome is for people to get jobs based on merit rather than ambition, and we should try to iron out the genetic bias towards male ambition.

Whichever view one takes, the conclusions are bound to be contentious. It could be argued that ambition attests to a desire to be successful, and if the desire is not there – whether or not it comes from a personal feeling or genetic disposition – then we should not attempt to rectify this difference. According to this view, women, on average, do not *want* the same level of success, so we should not try to level the field. Moreover, ambition and risk are difficult things to quantify. How do we measure the average difference, separate it from environmental effects such as upbringing and education, and then apply it to employment policy? An added complication is the possibility that success in some high-flying careers requires a positive ability and disposition to take risks. Would women have wanted to be Venetian explorers?

Of course they would. We should bear in mind the following: if there *is* a difference between the sexes, then the difference is statistical. It does not apply to every single person. Take mathematics, which is one area of the school curriculum in which, according to the figures, boys tend to do better than girls. One study showed that in any random pairing a boy will do better than a girl in a maths test 63 per cent of the time. That does not mean that all boys are better than all girls at maths. We must be careful of making such assumptions. If one is hiring for a job, and you assume that any given male candidate is better at maths than a female candidate, you will be wrong almost four times out of ten. Cultural differences, family influence and genetic differences between individuals will account for much of this variation. It is amazing, given the power and scope of society and upbringing, that evolution can be detected at all in our behaviour, but it can be, and it influences and underlies every single culture.

Competition has been a driving force in the evolution of mankind. Competition is not necessarily beneficial in the short term; indeed, it can make our lives exceedingly difficult. It can cause us to compete for the sake of competing, take risks with our money and our health; it even allows us to risk our lives for no apparent reason. Not only do some people buy flashy and expensive cars, they then proceed to drive them too fast and crash.

Evolution hurts. This is a theme that will figure strongly in the next chapter, in which we investigate the human instinct for violence. But in the meantime, we should bear in mind that competition can in some circumstances be an extremely positive force in human society, not necessarily, as Adam Smith might have said, for the efficient workings of a free market and the generation of wealth.

The instinct for competition plays a role in every branch of human endeavour, from the writing of symphonies to new scientific discoveries. Natural talent is rarely enough for the true geniuses and high achievers in any field. Talent is often accompanied by persistence, a willingness to take risks, and a burning desire to be the best.

chapter**six**

Violence

Lombroso

Cesare Lombroso was the most celebrated criminologist of his day. Born in Verona, and educated at some of continental Europe's most prestigious universities – Padua, Vienna and Paris – his academic career reached its apex in 1876 with the publication of his most famous work, *L'Uomo Delinquente*, or *The Criminal Man*. The book won copious praise in both lecture halls and courtrooms. Every urban centre in Europe was recording increased crime as industrialization grew apace and city populations sank into poverty, overcrowding and squalor. Criminality was a problem that had to be solved; Lombroso's theories provided a method to distinguish the born criminal from the law-abiding majority.

Lombroso tried to prove not only that criminality was hereditary, but that a criminal character was apparent in the physical shape of a person's head. His basic approach was really the sequel to the 'science' of phrenology, which had been highly popular in the eighteenth century. It was widely believed that different parts of the brain were highly specialized and that these different areas gave rise to different human attributes. When one part of the brain grew, it was thought, it would produce an indentation on the cranium which could be felt or even observed by a skilled phrenologist.

Consequently, a person's character could be predicted.

But Lombroso felt that tell-tale features in the face as well as the head gave crucial clues about the likelihood of antisocial behaviour. According to his theory, these 'stigmata', as he called them, were atavisms – physical traits reminiscent of earlier stages of human development which could reappear spontaneously after many generations of genetic invisibility. One particular feature to watch out for was apparently 'ears of unusual size, or occasionally very small, or standing out from the head as do those of the chimpanzee'. Murderers had 'prominent jaws' and pickpockets had 'long hands and scanty beards'.

By this reckoning, criminal psychopathology was akin to being under-evolved. Lombroso believed that ancient humans were much more violent and criminally minded than modern man. Modern criminals not only shared their ancestors' criminal tendencies, but also their physical appearance. They were living examples of the violent savage, a beast several rungs below modern man on the evolutionary ladder. The criminal man was essentially an evolutionary degenerate.

The theory was acclaimed as a bold application of Darwin's theory of evolution to the important political and legal question of criminality. Lombroso was thought of as a great authority and was even called upon to give evidence at criminal trials. In one such case, two brothers were on trial for killing a woman, but it was known that only one of them had committed the crime and neither would implicate the other. Lombroso examined both men and declared that one brother looked like a peaceful character and so was almost certainly innocent. The second brother, on the other hand, was a 'born criminal', according to the theory. Lombroso described to the court the man's enormous jaw, thin upper lip, unusually large head and enlarged frontal sinuses. This man had the physiognomy of the criminally minded, and on that basis he was duly convicted.

Had the theory gained enough support among the politicians and lawyers of nineteenth-century Europe, Lombroso's work might well have been adopted as a basis for eugenics. Criminals, viewed as unwanted genetic stock, could be cleansed from society by selective breeding or programmes of sterilization. Once the physical traits of criminals were properly classified, children who were unlucky enough to have been born with certain physical features that suggested a future life of crime could be treated before they had the chance to break the law. After some years, however, Lombroso's theory fell out of favour. The connection between facial features and a tendency towards criminal behaviour was shown to be entirely spurious. Nor are the sort of physical features Lombroso described true atavisms, which are actually extremely rare genetic throwbacks causing such physical conditions as extra nipples or 'werewolf syndrome', a condition that causes hair to grow over the entire body and face.

With Lombroso discredited, criminologists turned their attention to more sensible notions about the causes of criminality – upbringing and social and economic background. Atavisms, stigmata and strangely shaped skulls went the way of phrenology and other Victorian pseudo-science. We had entered the era of social science, and the focus turned towards the idea that criminals are made, not born. Social scientists portrayed criminality as having its roots in the breakdown in family relationships, cycles of violence, poor education and financial hardship rather than as a result of deterministic genetic programming. Still, however much Lombroso's theory now irks us, we should spare a thought for where the poor gentleman ended up. After his death in 1909, his head was pickled in a jar and has been on display ever since at the Museum of Criminal Anthropology in Turin.

As it happens, the idea that criminality is hereditary has recently resurfaced. The study of the human genome and our new-found abilities to trace the effects of particular genes have allowed researchers to test directly whether certain genes are implicated in violent behaviour. The subject remains a source of great controversy and has provoked the most extraordinary academic brawls.

The Violence Initiative

In the spring of 1992, rumours began to spread of a major American research programme funded by the National Institute of Mental

Health (NIMH), investigating the biological basis of violence. It was called the 'Violence Initiative'.

The director of the NIMH, an eminent psychiatrist called Frederick Goodwin, subsequently held a public meeting in Washington DC. Goodwin put forward the highly controversial idea that certain individuals in society were genetically predisposed towards violence. In other words, some of us are hard-wired for criminality, and the Violence Initiative intended to find those people who were. Goodwin drew comparisons with primates, including chimpanzees, among whom there are apparently certain individuals who also have a greater than average fondness for physical confrontation. Then, unwisely, he elaborated, comparing human criminals with 'hyper-aggressive' or 'hyper-sexual' monkeys. Goodwin even went on to say that calling crime-ridden urban America a 'jungle' might be more than just a metaphor.

That was a big mistake. Young black men commit a large proportion of crime in urban America. Their arrest rate for violent crimes, for instance, is six times higher than that of young white men. Goodwin's comparison with chimpanzees was bad enough, said the critics, but to imply they lived in a 'jungle' was beyond the pale. In his account of the ensuing row, Tom Wolfe wrote, 'That may have been the stupidest single word uttered by an American public official in the year 1992.' Congressmen, senators and scientists alike fell over themselves to condemn Goodwin and his research programme. They told him his remarks were racist and then took a swipe at his entire methodology; they said a primate study was 'a preposterous basis' for analysing the complexities of 'the crime and violence that plagues our country today'. In their view, Goodwin was a hair's breadth away from Nazi eugenics. The implication was that those people with the 'violence gene', if there is such a thing, could be found and somehow treated at birth. They raised the spectre of 'the mass drugging of helpless, innocent inner-city children'.

In the face of a storm of protest the NIMH backed down, first trying to deny the existence of the programme and then insisting there were no racial implications to the research. But the ball had already started to roll, and research continued, albeit discreetly. David Wasserman, a professor of law, attempted to breathe some life back into the programme by organizing a conference in an obscure seaside town on the Maryland coast. The National Institutes of Health, unwisely, partly funded the event, but only after Wasserman agreed to invite scholars who would speak against the notion of a genetic cause of crime. Nonetheless, the conference was invaded by a group of angry protesters who walked in shouting, 'Maryland conference, you can't hide – we know you're pushing genocide!' 'The present moment resembles that moment in the Middle Ages', Tom Wolfe later pointed out, 'when the Catholic Church forbade the dissection of human bodies, for fear that what was discovered inside might cast doubt on the Christian doctrine that God created man in his own image.'

Scientists have to pick their way carefully through the minefields of political taboos and cultural sensitivities. They must also be aware of the possible misuse of science, and it is quite right that credence should not be given to politically motivated or racist theories. But researchers also need to keep on doing 'science'; that is, they must continue impartially and objectively to hold up a mirror to the world around us, and that, of course, includes the human world. Despite its incendiary implications and the way it was mishandled, the Violence Initiative was a potentially interesting programme. As we'll see in this chapter, there are a number of tantalizing points of evidence which suggest that there are strong genetic and biological roots for violent inclinations.

Even before Lombroso's time there was evidence that certain regions of the human brain were intrinsically linked with aggression. Phineas Gage was a young railroad construction supervisor in Vermont. In September 1848, while setting up explosives for rock-blasting, he pushed a steel rod over one inch thick into a hole containing incendiary powder. There was a violent explosion that thrust the end of the steel bar into his face at a high speed. The bar pierced his left cheek just above the jaw, destroyed his left eye, went through the frontal part of the brain and made an exit through the top of the skull on the other side, landing some twenty-five yards away. Most of the front left side of Gage's brain was destroyed. He lost consciousness and started to have convulsions. However, he came round moments later and was taken to a local doctor, John

Harlow. To everybody's surprise, he was talking and could walk.

Gage survived the resulting infection and the blood loss, but he was a changed man. Once sensitive, intelligent and respectful, after the accident he became aggressive, impulsive and rude. We now know that a region at the front of the brain known as the prefrontal cortex was disrupted. It plays a key role in our emotional processing, and appears to have an important part in mediating an aggressive reaction.

Recent research projects have involved scanning brain activity in real time in order to analyse how violence or aggression 'appears' in a working brain. Is there a physical change or an unusual structural feature that causes one person to be more violent than another? While it doesn't appear that there is a 'violence centre' in our brain, the amygdala – the almond-shaped communications centre in the heart of the limbic system – does appear to initiate feelings of fear and aggression in much the same way as it would launch the fight-or-flight response in a dangerous situation. The prefrontal cortex is wired directly to the amygdala and seems to be the 'brake' on aggressive behaviour, reigning in and mediating the signals from the lower brain regions.

If that is the case, what is going on inside the brains of particularly violent individuals? Do they back up theories on the involvement of the amygdala and prefrontal cortex link? In one study, convicted murderers had their brains scanned and it was discovered that in the majority of cases the prefrontal cortex and sometimes also deeper brain areas such as the amygdala were functioning abnormally. In another study, researchers assessed a group of individuals who were known to have violent tendencies and found that their prefrontal areas were much smaller when compared to other normal individuals.

However, we must take care with such evidence. Are these clues a biological cause of violence or the consequence of some other unknown factor? For instance, there is strong evidence to suggest that other biological factors are at work. Hormone levels in our brain appear to affect levels of aggression. A high testosterone level seems to influence the intensity of one's aggressiveness, and these levels vary considerably between individuals. They also fluctuate significantly depending on your day-to-day circumstances. As we have seen, testosterone levels vary in line with general competitiveness. In one study of American men in different professions, the highest reported testosterone levels were found in trial lawyers, and the lowest among religious ministers. There's also the suggestion that a limited capacity to absorb serotonin, the primary indicator of depression, into the brain cells correlates with increased aggression. This implies that Prozac-like drugs could curb violent personalities. Other theories predict that individuals who are relatively insensitive to stimuli in the environment around them will seek out high-risk activities, including criminal ones, to increase their level of arousal.

Despite determined efforts to understand the neurological processes involved in violence, none of these projects has led to any solid conclusions. Even if there are differences we cannot be sure that the structural or functional quirk is not the *result* of an already existing disposition for violence. Development of the brain as a child proceeds, in part, according to the environment he or she grows up in – upbringing, relationships, emotional experience. Important physical changes in the brain can depend on external stimuli, just as violin-playing causes the part of the brain that controls left-hand dexterity to grow. One person's brain may be physically different, but that does not mean it is due to an inherited trait; differences in the connections may be the after-effects of an outside influence. So we land squarely back at the sociological explanations of violence.

The task of separating the environmental and the biological is notoriously difficult. We do not yet understand how genes, brain chemistry and neural structures interact, partly because we are in the dark as to which is the cause and which the effect. Interestingly, though, brain imaging may have an impact on crime and punishment in a more direct way. A commercial company in the US claims it has developed a brain imaging lie-detector. The suspect is shown words or images connected to the crime. If the suspect recognizes certain details of the case – for example the victim's appearance or clothes, or the location of the crime – then the system detects a particular kind of neural response called a 'mermer' that indicates recognition, and therefore implies guilt. An Iowa court recently allowed this evidence to be used in the defence of a man charged with suspected murder, although it didn't persuade anyone of his innocence.

For the modern-day Lombrosos who set out to catch the violent criminal before he or she has committed a crime, science is clearly not yet ready. No neuroscientist, policeman or parent can tell whether the smiling face of a newborn baby hides the mind of a potential killer. Neither a funny-shaped skull nor a particular neural quirk will predict his or her future behaviour; the interplay of environment, genes and biochemical make-up is too complex and opaque.

An instinct for violence?

Violence comes in a bewildering array of forms: infanticide; lethal clashes of males competing for females (and occasionally vice versa); gang-related violence, organized warfare and slavery; rape and other forms of sexual battery; battles over territory. And believe it or not, these are just examples from non-human species.

Clashes between animals (of the same species) are often benign, especially when the rewards are temporary or insignificant. Speckled-wood butterflies competing for warming spots of sun on the forest floor never get injured; they can find another shaft of sunlight easily enough. Fights over mates may be more serious, but it depends whether the animal in question has only one chance to mate in his lifetime. Only 2 per cent of male red deer are seriously injured in their antler-rattling contests. When the stakes are higher, though, there is more of a chance of violence. One long-term study found that 25 per cent of a wolf population was killed in confrontations with other wolves. Gall insect larvae battle over access to the precious plant galls they occupy, and six times out of ten one of the combatants will end up dead. If two honeybee queens are competing for control of a colony, a crucial struggle which will determine complete genetic success or failure, they will fight to the death, every time.

Why should *Homo sapiens* be any different? As with our instincts for sexual reproduction, competition and survival, a human instinct for violence will have been carved from generations of natural selection in our ancestral homeland. The savannah provided only so much food in the form of fruit, meat, roots and other necessities of the palaeolithic diet. What's more, our early human ancestors spent thousands of years having to hunt and kill their prey. Territory, too, has always been a precious resource; the best vantage points, hideouts and water holes would surely have been a source of dispute. Sex was never likely to be a civilized, peaceful activity, and the competition for the most fertile and best-looking mates must have led to fierce clashes. It's obvious that early humans had to be violent, and it follows that violence must be programmed into our genes. Doesn't it?

Hobbes and the state of nature

Political philosophers have long been intrigued by this question. Society, states, government and law all have ways and means of constraining and controlling human behaviour, so early human life, when these controls did not exist, may tell us a little about the kind of political system modern humans instinctively want, or need to best control us. Thomas Hobbes, the greatest English philosopher of his generation, was one of the first to grapple with this question. In 1651 he wrote his most famous work, *Leviathan*. The 'Leviathan' of the title was an absolute ruler, a sovereign whose power reached into every corner of civic life. It was also a solution, Hobbes concluded, to the question of the justification of a political system and all it implies – government, taxation and law.

Hobbes argued that in order to have a theory of politics in a modern civilization, one must imagine the condition of mankind before politics and organized society existed. This period he called the State of Nature. Famously, he painted a despondent picture of life in the State of Nature; there was, he said, a lack of creature comforts, of arts, of letters, of industry of any kind, and the worst part was the 'continuall feare, and danger of violent death; and the life of man, solitary, poore, nasty, brutish and short'.

It was a pessimistic version of events, melodramatic perhaps, but in the context of the political situation in which Hobbes lived, understandable. He had survived the English Civil War and believed that warfare was the worst of all possible evils; in a time of war we act to preserve ourselves, so all other moral considerations must take a back seat. The thrust of his argument is that in the State of Nature we were so irredeemably amoral, so lacking in any sense of right and wrong, that it would have been in our interests to make an agreement. He called this agreement the Social Contract.

The Social Contract symbolizes the fact that co-operation was the only way out of the mess. It is a product of free individuals who decide on a rational form of politics and communal life, a state that Hobbes called the Commonwealth. The State of Nature was so brutal, so ferocious, that the only solution would be to create a Leviathan of immense power and total control.

Hobbes was a brave and brilliant thinker, but his theory was built on rocky ground. The State of Nature is supposed to be devoid of all 'civilizing' influences and social pressures, but it seems unlikely there was ever a time when humans were entirely non-social, solitary hunters or scavengers who only came together to breed. Indeed, human life is defined by group living and social interaction. Co-operation, as we shall see in the following chapter, is an enduring fixture of human life, even when there are no laws to enforce it. Hobbes's State of Nature was too abstract and far removed from reality; even as a thought experiment it fails to cast any light on human nature.

Yet the underlying question is still valid. Where does violence come from? Is it a product of human culture or does it have evolutionary roots? Are we capable of being equally warlike and peaceable, depending on our cultural traditions and individual predispositions?

One view held by scientists

In 1986, a group of twenty scientists gathered in Seville under the auspices of UNESCO to mark the International Year of Peace. The purpose: to discuss the causes of violence and warfare and draw up a manifesto on the subject. The scholars were drawn from a number of disciplines including psychology, sociology, anthropology, neuroscience and zoology. One would think that agreement on such a contentious issue would be elusive, but surprisingly, given their divergent backgrounds, they managed to create enough of a consensus to issue a very clear statement. It began:

> *IT IS SCIENTIFICALLY INCORRECT to say that we have inherited a tendency to make war from our animal ancestors.*

This was certainly a bold start. Categorizing any belief as 'scientifically incorrect' is a confident gambit, and a surefire way to raise the hackles of anyone who disagrees. The thrust of the opening paragraphs was that *Homo sapiens* is the only species that practises organized warfare regularly using tools. That much was fairly uncontroversial (although, as we shall see, not entirely true). The statement continued:

> *War is biologically possible, but it is not inevitable, as evidenced by its variation in occurrence and nature over time and space. There are cultures which have not engaged in war for centuries, and there are cultures which have engaged in war frequently at some times and not at others.*

This is undoubtedly true. Warfare is not a regular, predictable feature of human life. Cultural factors clearly affect the level of violence in any society. They went on:

> *IT IS SCIENTIFICALLY INCORRECT to say that war or any other violent behaviour is genetically programmed into our human nature . . . Except for rare pathologies, the genes do not produce individuals necessarily predisposed to violence.*

And if that was not clear enough:

> *IT IS SCIENTIFICALLY INCORRECT to say that in the course of human evolution there has been a selection for aggressive behaviour more than for other kinds of*

> *behaviour. In all well-studied species, status within the group is achieved by the ability to co-operate and to fulfil social functions relevant to the structure of that group.*

In other words, humans are not instinctually aggressive; we are co-operative, altruistic and sociable. And just for good measure, the neuroscientists added:

> *IT IS SCIENTIFICALLY INCORRECT to say that humans have a 'violent brain'. How we act is shaped by how we have been conditioned and socialized. There is nothing in our neurophysiology that compels us to react violently.*

In conclusion, they stated:

> *. . . biology does not condemn humanity to war . . . The same species who invented war is capable of inventing peace. The responsibility lies with each of us.*

With that, the Seville Twenty staked their claim to the 'environmental' explanations as being the all-powerful causes of violence and warfare. Their (admittedly admirable) intentions were founded on a fear of misuse of scientific theory, particularly evolutionary theory; the ghost of Social Darwinism, used to justify the most horrendous acts of war and violence, including Nazi genocide, has still not been laid to rest. They believed passionately that evolutionary explanations of violence had created a culture of pessimism; if violence is in our genes, they reasoned, people would resign themselves to a violent and war-torn world.

To the casual reader, it appeared as though the Seville statement was a discovery. It was, of course, a particular point of view thrashed out by a particular group of academics based on the available research up until that point, but the language they used indicated a breakthrough in the human sciences. Violence, they were saying, is not genetic. We may be culturally violent, but underneath we are peaceful beings. Of this it seemed quite clear they were in no doubt.

It would, I suppose, be nice if they were right. The statement, in many respects, outlines a sensible position in that it denies that genetic programming is an all-powerful force that dominates our lives. Our personal and social lives *are* governed by a constant flux of cultural forces – political culture, upbringing, social and economic situation. It would be absurd to think otherwise, and their implicit characterization of evolutionary theorists as people who would deny this fact is, in any case, a caricature.

But to their shame, the statements of these scientists were made on purely subjective grounds. Well-meaning intentions lay behind the Seville statement, but the Twenty did not add any real evidence in support of their pronouncements. The oft-repeated phrase IT IS SCIENTIFICALLY INCORRECT came across as an attempt to persuade us that two plus two equals five, because, to be fair, it would be better all around if these statements were true. But the truth is that a major challenge still exists: to discover, using the blunt and imperfect tools of the various human sciences, whether or not there is a genetic component to violent behaviour; and if there is, to what extent it exists and whether it can be modulated by a person's environment. This is a critical question of enormous importance. Politicians, sociologists, policemen, psychiatrists, prison wardens and everyone who deals with any aspect of crime and violence are still awaiting the answers. Which is why the scientific community is continuing to take up the challenge and seek the biological roots of violence.

Why be violent?

The idea that a disposition to violence is a pathology has been around for many years. Pathologies describe conditions in which some aspect of the human mind or body has ceased to work properly. Liver disease is a pathology, as is schizophrenia, which is accompanied by certain biochemical changes in the brain. Perhaps, then, violence is a symptom of a mental disorder, a type of behaviour that shows a failing in our mental processes or abilities.

To a certain extent this view is reflected in modern psychiatry. Psychiatrists believe that many persistent criminals suffer from a condition called anti-social personality disorder, or ASPD, which is essentially a milder form of what's known as psychopathy. The outward signs of ASPD include impulsiveness, recklessness, a habit of getting into fights, deceitfulness, an inability to conform to social norms and a lack of remorse when crimes have been committed. They are traits that wouldn't be hard to find among the inhabitants of the nation's prisons.

Most psychiatrists restrict their work to the living, but Dr Eric Altschuler of the University of California believes that Samson, the biblical war hero whose strength resided in his long hair, also suffered from ASPD. According to the Bible, Samson vowed to serve God and the Children of Israel, and the best way he could do that was to fight. He was a feared warrior and a great leader of men. On one day alone he killed a thousand Philistines single-handedly, using the jawbone of a donkey as his weapon. Altschuler believes, however, that Samson's heroic reputation may have to be revised. He says that had Samson carried out such feats in our own age, he'd be regarded as a 'bit of a thug'; he was a bully, and showed no remorse for his violence. In divulging his secret to Delilah, after she had tried to have him killed at least three times, he revealed himself to be impulsive and had little regard for his own safety. Samson also broke his Nazarite vows – he was not supposed to drink wine, cut his hair or touch dead bodies (although he did eventually make up for his transgressions by destroying a temple, pulling down the pillars that supported the roof and killing himself along with three thousand Philistines).

Samson was one of the very few biblical characters to commit suicide, and it is probably fair to say that his relationship with Delilah was what we would call 'dysfunctional', but did he really have a mental illness? To bring the argument up to date, does a violent and thuggish general of a conquering army, or for that matter the ring-leader of a gang of football hooligans, have a pathological condition?

The suggestion that a disposition to violence is pathological has two implications: firstly, that it may somehow be cured, and secondly, that violence had no useful role to play in our evolutionary past. But if violence could actually be seen as an adaptive form of behaviour, it would make no sense to think of it as a pathology. (That is not to say that violence is a desirable feature of modern life; in the modern context, violence could certainly be seen as maladaptive.)

Let us consider the situations in which people become violent. These usually occur when something important to them is at stake, as with the red deer or the queen bees. Territory, mates and food – these were the basic and essential commodities for life on the savannah. Now, imagine what might have happened during a conflict over a water hole on the savannah, or for that matter in a confrontation over who gets a parking space in a supermarket car-park. Intuitively, it seems as though violence is not a conscious tactic. It's usually an immediate and unconscious reaction, there to be used as a last resort after an attempt has been made to reason with the interloper. And it turns out that our ancient fight-or-flight instinct is not only initiated by feelings of fear, as we saw in the first chapter, but also by feelings of anger. The flow of adrenalin and rising blood pressure are necessary conditions for engaging in violence (or for running away from a violent opponent). So if our anger reaction is so similar to our fear response, does that mean we are 'built' to react with physical aggression, or to expect aggression in others?

Most of us are not physically built for violence. From *Australopithecus* onwards we have become much more frail; modern humans are weak, substantially weaker than an average adult chimpanzee, and we lack the teeth and claws which make many species effective aggressors. But maybe our ability to use tools made our physical limitations irrelevant. We can swing a club, throw a spear or fire an arrow instead. Weapons, the first of which was probably a stick or a stone found on the ground, have become more and more lethal as our bodies have become weaker. Who needs the wrath or despair of Samson when you've a nuclear warhead?

In tribal or hunter-gatherer societies with no central authority, violence is certainly not considered to be pathological. In fact, the most successful members of the group in terms of wealth, status or number of children are often those who engage in the most violence against others within the group and outside the group. Aggression grants kudos, a precious commodity in any society.

In the modern post-industrial West, the most likely candidate to commit a violent crime is, statistically, the inner-city poor black youth. This is not because poor black youths are genetically predisposed to violence, although some theorists have tried unsuccessfully to argue that case. It is because the social conditions in which they live are the most conducive to crime, both in terms of motivation and in terms of opportunity. The fact is that the poor, the frustrated and the dispossessed are much more likely to take risks. They simply have less to lose. As Daly and Wilson say, 'a reason to doubt that one will be alive tomorrow is reason to grab what one can today'.

Lack of education, poverty and a culture of violence within the community – these are powerful predictors of, and triggers for, crime. But that does not imply there is no room for a genetic explanation. There is bound to be a genetic variation in predisposition to violence, just as there is variation in height, in intelligence and in a liking for vanilla milkshakes. But it's important to point out that the very latest genetic studies prove that differences in our genes do not match our perceived notions of different races in any way. There is, in fact, more genetic difference overall between any two random individuals than between two people from different so-called 'races'. As such, the high likelihood of an impoverished black youth committing a violent crime says far more about the effect of social conditions than about the genes of black people.

But glance at any crime statistics and you'll see one glaring fact jumping off the page. It turns out that there *is* one particular group who have a definite genetic marker for violence – males. The vast majority of crime, and violent crime in particular, is committed by men. As with our sexual instincts, this difference can almost certainly be pinned down to genetic differences between the sexes. Boys will be boys, and it seems likely that selection has provided men with a greater predilection for physical aggression, and the physical strength to use it.

The violent male

Margo Daly and Martin Wilson have carried out a fascinating study of crime statistics across several dozen different cultures ranging from an English university town to the BaSoga people of Uganda. They drew out the figures for 'same-sex' murders – murders which involved males killing males, or females killing females. Male-male murders made up the vast majority of cases, anywhere from 85 per cent to 100 per cent, with most cultures scoring in the mid-nineties.

Each case would obviously be a story in itself, but we can imagine some likely scenarios. The two men may have been rivals in business or crime, or partners, one of whom has been betrayed by the other. The murders could be the result of bar brawls between two strangers that start off as pushing and shoving and develop into serious violence – often a clash over status or an attempt to save face. Fighting over women is not uncommon. Violence could also result from arguments over unpaid debts or deals that went wrong. 'Gold', wrote Shakespeare, 'is worse poison to a man's soul, doing more murders in this loathsome world than any mortal drug.'

Of cases that involve men killing women, most of them are husbands and wives or boyfriends and girlfriends. Domestic murders – and nine times out of ten these involve men killing women – are mainly the result of sexual jealousy or suspected infidelity. We've already seen the raw power of the jealousy instinct, and how quickly this can spill over into violence and bloodshed.

Boys will be boys

It used to be a truth universally acknowledged that boys like to play with guns whereas girls prefer to play with dolls, but for many years there have been attempts to undermine these hackneyed stereotypes in favour of a more equitable view of children's gender identity. Studies have shown that today's parents are as likely to buy their child a toy which does not match its traditional gender preference as one that does. However, the same studies have shown that when it's the child who chooses the toy for him- or herself, there is a very high chance that he or she will pick a toy that would have been traditionally assigned to their gender.

Recently, the major chain of toy stores Toys R Us decided to buck the liberal trend and announced plans to redesign the layout of its stores so that products were grouped according to whether they were likely to be bought by the same customers. The result was a complete separation of the stores into boys' toys and girls' toys. Barbie dolls and Glitter Girl Make-up were now several aisles away from semi-automatic pellet guns and Monster Trucks. The company claimed that their own research showed gender differences start as young as two years old; 'In general terms,' they said, 'girls are more interested in entertainment that is relationship-oriented,' and boys are more 'action-orientated'.*

Toys R Us were actually reflecting an increasingly popular view among evolutionary psychologists. Sexual selection is a powerful force; in fact, as we've seen, it was how genetic differences between the sexes come about. Males have to compete, and competition involves aggression, action and pointlessly dangerous activities. And there is another, more practical reason for a non-violent mentality to become a more common female trait. Given the need to suckle their infants (probably for a much longer period than is usual in modern society), women would have spent more time with their children; they took on the role of primary care-giver. We can speculate that women who were more empathetic, caring and less prone to violence and aggression would have made a better job of child-rearing.

Daddy's girls

But there's one small group of females around the world who don't follow that typically female trend of being non-aggressive. In fact, they are quite the opposite. A groundbreaking British study of their genes has recently revealed a fascinating insight into how the 'violent male' tendency is created in all men. Two chromosomes determine our sex. All eggs contain just one X chromosome, while sperm can carry either an X or a Y chromosome. On fertilization, fusion will occur, and all normal embryos will receive either an XX (and so become female) or an XY (and so become male) as their twenty-third chromosomal pair. So everyone has at least one X

* They were, however, forced to reverse their plans after (adult) members of the public complained that they were enforcing gender stereotypes.

chromosome, and women have one X from each parent. For males, the X chromosome has to come from the mother. And, as we shall see, the X chromosome plays host to genes that have a huge effect on one's behaviour and personality.

Turner's Syndrome is an uncommon genetic disorder that afflicts one in two thousand girls born each year. It is caused by a missing X chromosome; their single X chromosome can be inherited from either the father or the mother, so instead of being XX they are in fact XO. In many respects human development is resilient to this crude error. The loss of a single X chromosome, unlike the loss of one of a pair of most other chromosomes, is not necessarily lethal. It is true that a number of embryos with just one X do not survive to the end of pregnancy, but girls who are born with Turner's Syndrome are often relatively normal. They tend to be of short stature, have a thicker neck than normal and slightly bent elbows, and they generally suffer from infertility because their ovaries are not properly formed. Still, many girls with Turner's have great difficulty learning social skills, and they tend to be more disruptive, aggressive and generally anti-social. They butt into conversations, they misread people's facial expressions and body language and they're often very insensitive.

In fact, when it comes to social skills, these girls often tend to act quite like badly behaved boys. David Skuse of the Institute of Child Health in London discovered that these personality traits depend on which X chromosome the girls inherit. The girls who end up with their X chromosome from their mother are significantly more disruptive and have greater problems adjusting in a social environment than the girls with their X chromosome from a sperm. Parents of Turner's and non-Turner's children reported back using a scoring system of anti-social behaviour which ran from one (as nice as pie) to twenty-four (psychopathic and potentially murderous). Girls with a normal complement of two X chromosomes scored an average of two points. Boys scored an average of four, as did the Turner's girls with the paternal X chromosome. But the Turner's girls with a maternal X scored a relatively high nine on the anti-social scale.

The researchers believe that there are genes responsible for modulating behaviour on the X chromosome, and those genes that cause anti-social behaviour are imprinted or 'highlighted', and thus more active, on the maternal X chromosome. In a girl with a normal complement of chromosomes, any anti-social tendencies would be countered by genes on the paternal X; but Turner's girls with the maternal X are lacking that brake on their behaviour.

What does this tell us about the role of the X chromosome in normally developed children? Remember that most girls have the maternal *and* the paternal X, whereas *all boys* have just the maternal X chromosome. Boys, therefore, have the anti-social maternal genes without the paternal 'brake'. In other words, most girls are nicer than most boys, and the reason is genetic. Men are programmed to behave badly.

According to the theory, boys should therefore be more disruptive, insensitive and anti-social than girls, and most parents and teachers will tell you that this is in fact the case. It also applies to displays of violence and aggression. A study of pre-school children showed that during confrontations between two children of the same sex, boys are more than twice as likely to use 'heavy-handed persuasion', defined as physical force or threats, than more peaceful forms of conducting a dispute. In almost all cases, girls are far more likely to try to talk their way out of a confrontation rather than use physical force.

Of course, a child is not just a pre-programmed bundle of genes. Upbringing and socialization have a great deal to do with behaviour. Boys are many times more likely than girls to have learned aggressiveness from their parents and other people around them who may have influence. Girls do play with dolls, and boys with guns, and Barbie never toted an M16 machine gun. It is always difficult to separate nature and nurture, but the Turner's Syndrome study does exactly that; on average, there are unlikely to be statistically significant differences between the upbringing of a girl with the maternal X and a girl with the paternal X. Their differences on the anti-social scale are therefore clearly a product of the genes, and from this we can infer that the same genes affect boys in a similar way. The 'X-factor' (no pun intended) seems highly likely to be an important component in the behavioural differences between boys and girls, and also between grown-up men and women.

Other genes

Just nine years after the declaration of the Seville Twenty, there was an important meeting in London which took a much more serious look at the influence of genes on human violence. The International Ciba Symposium on Genetics of Criminal and Antisocial Behaviour gathered many of the leading names in the field and strong evidence was presented suggesting that there were a number of genes which may be implicated in violent behaviour. Members of the symposium were not prepared to be merely politically correct, but as Sir Michael Rutter, the distinguished child psychiatrist who opened the meeting, said, participants needed to be clear about the risks of exaggerating any genetic influences or producing discriminatory labelling.

One of the first bits of genetic evidence which was relatively compelling was the simple recognition that different laboratory strains of mice, differing only in a few of their genes, showed considerable differences in their penchant for aggressive behaviour, and some mice missing a particular gene on mouse chromosome 9 (the equivalent gene is on chromosome 6 in humans) are of great interest. This gene produces certain brain receptors, in particular receptors for serotonin. Mice without the gene behaved differently. Though they appeared completely normal when exploring new areas in a fresh cage, and bred and ate like other mice in a completely docile way, they acted very aggressively when confronted with new mice they hadn't previously encountered.

At the same scientific meeting, Finnish researchers presented quite strong evidence that changes in the equivalent gene in humans was associated with anti-social behaviour, a violent nature and a tendency to alcoholism. At present, studies are being extended to see to what degree this gene varies in its structure in the general population, and in those with behavioural problems.

A somewhat related gene is that which produces a substance called monoamine oxidase A (MAOA), which affects neurotransmission in the brain. In some families this gene is abnormal, and in one Dutch family the men who are affected show increased impulsive behaviour, aggressive sexuality and a tendency to commit arson. But the scientists who were working on this gene were at pains to point out that it was unlikely the MAOA gene was, in itself, an 'aggression gene'. The complexity of variation in the behaviour they observed and the highly multi-faceted effects of deficiency of MAOA on neurotransmission make a direct causal link very unlikely. So the jury is still very much out on genes that contribute to violence, and there is a long way to go before we can identify certain genes associated with aggression that would have had a selective advantage during our time on the savannah.

One of the most tantalizing aspects of the London meeting was the paucity of reports on twins studies. Twins are of remarkable use to geneticists. Identical twins have, of course, identical genes. Non-identical twins have the same chance differences in their genes as do any brother or sister, but they will have been exposed to a more similar environment before birth by sharing the womb at the same time. Moreover, after birth the environmental influences will tend to be rather more similar, more so than for siblings who have even a slight birth age difference. Studies of these various groups using careful statistical methods and measurement can give an insight into the relative influence of genes and the environment. Studies of this sort can be extended and can have several permutations. There are a number of cases where twins, even identical twins, have been separated at birth, usually for adoption into totally unrelated families – each with, of course, a different environment. Studies of such twins have provided much information on the influence of genes against that of nurture.

There is still a shortage of information, but one study by Dr Michael Lyons, a psychiatrist from Boston, showed some interesting and controversial trends. He studied 3,226 twin pairs, all male, who had served in the army during the Vietnam period. Roughly half of the twin pairs were identical. He looked in particular at histories of arrest and criminal behaviour. It seemed possible from his study that a common environment was more influential in criminal behaviour before the age of fifteen, but criminal behaviour after fifteen was more likely to be due to genetic influences. The shared environmental influence, of course, is less likely to be an issue once a child leaves its family unit, and there were considerable difficulties when it came to interpreting Dr Lyons' findings, not least because of the way in which the twins were recruited for study and the possibility that the sample was biased.

Women at war

But it is not only men who are violent and aggressive. A national holiday in Vietnam celebrates the suicide of two sisters, Trung Trae and Trung Nhi, who lived in the first century AD. These two legendary women, symbols of Vietnamese national pride, organized a revolt against Chinese rule which involved their heroically commanding an army of eighty thousand peasants. They gained control of dozens of towns, almost forcing the Chinese out of Vietnam for the first time in a thousand years, but the Chinese, with an army several times the size of the Trung sisters' force, finally overwhelmed them. They killed themselves before they could be taken prisoner.

Britain, too, has had its women warriors. It was not uncommon for ancient Celtic women to govern tribes and lead armies into battle. The most famous warrior queen was Boudicca who led the Iceni of Norfolk against the Roman army and at one stage sacked and burned London. Another Celtic Queen was Castimandua, the powerful and shrewd leader of the Brigantinians during the time of Emperor Claudius. When a group of Castimandua's soldiers was brought to Rome as prisoners, they bowed in front of Agrippina the Younger, Claudius's wife, assuming that she was the emperor.

But of all history's women warriors, the most famous are the Amazons. Their greatest queen, Myrene, conquered great swathes of the Middle East and the Mediterranean from Samothrace to Syria. It's said that in a battle in North Africa she once commanded an army of thirty thousand women on horseback.

Modern armies, though, have rarely allowed women to be front-line combat troops, on the assumption that (a) women lack the strength, stamina and physical prowess to match the performance of men; (b) women would not *want* to fight on the front line; and (c) the presence of women would prove to be disruptive, and dangerous for morale.

The first of these reasons is dubious. These days warfare is much less dependent on physical strength. There is no need to swing a broadsword or hump a cannon over miles of rough terrain, and there are now many front-line roles that do not require SAS-style levels of strength and speed. The second assumption is also suspect. The fact is that many women enter the services, and I imagine that some, if not most, would want to be part of a front-line force. The third reason, however, may be more troubling for the prospect of mixed front-line units. Women could well be the cause of divisions and conflicts between male members of the unit, just as men could be disruptive in a female unit. But I think in the heat of battle, jealousies or sexual rivalries would pale into insignificance.

In general, many women have shown themselves to be just as warlike as their male counterparts, and if they don't fight themselves, women can be the cheerleaders for warlike activity, even when the men aren't so eager. One eighteenth-century anthropologist described how women of the Ba-Huana tribe of the Congo used to make fun of their men if they didn't retaliate after an attack by a rival tribe. 'You are afraid,' they would say, 'and we will have no more intercourse with you.' The men, then, had little choice but to go out and fight. 'However unpalatable the fact,' says the military historian Martin van Creveld, 'the real reason why we have wars is that men like fighting, and women like those men who are prepared to fight on their behalf.'

My own laboratory is dominated by women: the ratio to men is about 5:1. And, of course, all there is sweetness and light. Nevertheless, I do not think we should underestimate the female instinct for aggression and violence. In a context of war between tribes, chiefdoms or nations – a phenomenon we shall investigate later in the chapter – female bloodlust may well match that of their husbands, brothers or sons. No-one would deny that with, on average, twenty times more testosterone pumping through his veins a male is more disposed, both physically and mentally, to being violent, but women are often more than willing to cheer them on – and, as history has shown, they're often willing to fight for themselves.

Going berserk

Although certain individuals are genetically more inclined to violence than others, there's another way to look at violence in society. Human beings have a capacity to *use* violence in certain situations, a strategy which might, historically, have been useful in acquiring mates, food, territory, status or revenge. Imagining violence as a strategy allows us to investigate its evolutionary potential as a sustainable form of behaviour; in other words, it's how we can work out whether evolution would have favoured violence among groups of early humans.

Recently, an inventive study has taken as its subject matter the race of people who are best known for fearsome violence, rape and pillage – the Vikings. Robin Dunbar and his colleagues analysed *Njal's Saga*, an Icelandic narrative that describes the lives and fates of ten Viking families.* Three of the families in the *Saga* are particularly interesting; each had within its ranks a man known to be a 'berserker'.

A berserker was a warrior known for his fearlessness, his bloodlust, his superhuman strength and high threshold of pain. Some people believe hallucinogenic mushrooms might have induced this feeling of invincibility; others think vast quantities of alcohol might have done the trick. Whatever their drug of choice, in the heat of battle berserkers showed a frenzied and violent devotion to victory and so became heroes; in fact, they were later employed as elite 'shock-troops' in the king's army. No doubt the Vikings' reputation as warriors is largely based upon this small minority, men who left an indelible impression on the unsuspecting foreigners who had the misfortune to meet them.

During peacetime, however, berserkers were demonized by their

* The sagas were written in the thirteen century and describe events which took place several hundred years before that date. Therefore, the accounts bear all the misunderstandings and creativity of a game of Chinese whispers. However, over the past century archaeologists have followed the descriptions of some of the burial sites and settlements mentioned in the sagas, and found that these sites did indeed exist. In addition, it is believed that the primary reason for writing the sagas was to record family history, so the portrayal of the lives of family members may be reasonably accurate.

own people. Violence was their *raison d'être* and they didn't stop just because they were no longer at war. Rape, murder and violence were all attributed to marauding (and presumably drunk) berserkers, and as such they became the bogeymen of Nordic peacetime life. Their family histories, however, allow us an interesting window into the strategic benefits of choosing violence over non-violence. Does it pay, from the point of view of evolution and the number of offspring one leaves behind, to be a berserker?

The answer turns on a point of Viking law. When a person was killed in peacetime, the family of the victim could either seek revenge and kill a member of the murderer's family or they could ask for compensation, known as blood money. Many such murders and their consequences are described in the sagas, and it seems that when a berserker committed a murder, the family would overwhelmingly choose blood money rather than a revenge killing. The alternative was far too risky, given the volatile state of the berserker. Conversely, if the killer were a non-berserker, more often than not the family would choose revenge. So the question is: how many surviving descendants did the berserkers leave compared with the others? Was 'berserking' a useful strategy?

The sagas show that despite their dangerous lifestyle – they were often killed in battle – berserkers did indeed tend to leave more relatives than non-berserkers. Overall, the strategy was successful. Using this rather extreme example, the research gives us an insight into why violence may arise, even in a population that is by and large inclined towards peace.

A more stable strategy

For a long time violence was considered to be an aberration, an act that was plainly non-adaptive. This was a legacy of group selection, which logically could not conceive of violence between animals of the same species to be adaptive. Any trait, such as a quick temper, that would cause injury or death, and therefore reduce the fitness of the species, would surely have been selected out. But with the lessening influence of the theory of group selection we have lost the naive and cosy view of non-violence being our 'natural state'. Hobbes was right, in so far as his State of Nature could never have been a peaceful idyll.

Strategy is all, over evolutionary time. That is not to suggest we make a conscious, rational decision whether or not to use violence according to circumstances. I use the word 'strategy' to mean a long-term evolutionary tactic, just as the adoption of bipedalism or a fear of snakes was a successful tactic over evolutionary time. We need a method of investigating whether we should, in theory, adopt violence as a tactic, assuming that the long-term overall aim is to maximize our genetic fitness. The evolutionary success of individuals will determine whether or not the tactics they adopt will then go on to survive and flourish. But of course, tactics depend on what everyone else decides to do too. In a football match, a certain formation of forwards and midfield players can be successful against one team but hopeless against the next. How does one 'work out' which behaviour is most successful in a population of others who may be behaving differently?

In the early 1970s, the famous evolutionary biologist John Maynard Smith set out to tackle this very problem, borrowing a technique used by mathematicians, economists and military analysts – game theory. Game theory was best known for its use in 'war games', or simulations of military conflict. When, after the end of the Second World War, it became clear that the Soviet Union had built the atom bomb, military analysts scrutinized the strategies open to the West. The possibility of Armageddon presented a rather thorny problem, but eventually everyone realized that mutually assured destruction, or MAD, was the optimum solution. MAD entailed pointing an enormous number of nuclear warheads at your enemy; everyone would die, no matter who fired the first missile. (Game theory does, however, depend on your opponent acting rationally. Had Stalin not died in 1953 and continued to become confused and delusional, things might well have worked out quite differently.)

Surprisingly, game theory also proved to be an extremely useful way of modelling the evolution of animal behaviour. Such models are bound to be simplified; there is no pretence at simulating real-life situations with all the nuances and complexities they entail. Instead, game theory presents a stripped-down, quantitative analysis – an approximation that can inform our theories of animal behaviour. It has produced some very interesting results.

Maynard Smith thought up a game in which two kinds of animal are competing for territory. This is a situation that could be found among thousands of different species in the real world, but our version is simplified and the rules are as follows. There are two kinds of player, the Hawks and the Doves. The first kind, the Hawks, are willing to fight for the prize, which in this case is territory. The Doves, as their name suggests, are peace-loving and cowardly.

I should emphasize here that these are not two different types of animal, they are two different strategies. Therefore no animal can know whether its opponent is a Hawk or a Dove before the fight, in the same way as you may not know if someone is a berserker when he's walking down the street.

Maynard Smith then assigned numerical values to the rewards and costs involved in the game. The winner of the territory gets +50 points and the loser gets zero points. If one of the players is injured, that will cost them -100 points. If one of the players begins a confrontation and then walks away, as Doves are prone to do, the cost of the confrontation is -10 points, whether or not they win or lose the territory. (This cost represents the time and energy involved in the ritual, and in the animal world one often sees two animals facing off, baring teeth, brandishing antlers or making threatening noises.) If a Hawk squares up against a Hawk, there will always be a fight. We can assume that all Hawks are equivalent in strength and speed, so each has a 50 per cent chance of winning. The winner, therefore, will get the territory – +50 points; the loser, who is injured, is deducted 100 points. If a Hawk confronts a Dove, the Dove will always walk away and let the Hawk win. The payoff for the Hawk will be +50. The Dove has to bear the costs of the face-off and is charged -10 points. No-one is injured. If two Doves challenge each other, neither will get around to fighting. Again, each Dove has to pay -10 points for the face-off, and the chances of winning are again fifty-fifty. Therefore, one Dove will get 40 points, and the losing Dove will be charged -10.

So, what happens if we begin with a population made up entirely of the peace-loving Doves? Each fight will be decided non-violently after a face-off. If the game is repeated many times, with all members of the population fighting random opponents, the average payoff for each player will be the total payoff per fight, 40 minus 10 divided by two (the number of players), which equals +15.

So far, so good. The average payoff is positive, which bodes well for the Doves. But what if we bring into play the rule that every so often there will be a mutation, and one player will switch strategy? What happens if one solitary Dove changes into a Hawk? The Hawk will obviously win every fight, gaining 50 points per confrontation, compared to the Doves' average of +15. This numerical success, which can be thought of as its evolutionary fitness, means that the solitary Hawk will thrive, constantly winning fights, allowing it to live long and prosper. Hawk genes will therefore spread through the population at the expense of Dove genes.

Now, suppose the population consists entirely of Hawks. Every single fight will end in violent conflict. The average payoff for every player, in the long term, is -25 (the total payoff per fight is -50, divided by two). In this environment, a solitary Dove will do well: the Dove's average payoff will be -10 (it will walk away every time from a Hawk), which is substantially better than the Hawk's average of -25. Therefore, the Dove's genes will prosper and spread at the expense of the Hawk genes.

The bottom line is that neither an all-Hawk population nor an all-Dove population is stable. Maynard Smith translated the game into the equations of game theory and realized that the only stable populations would have to be a *mixture* of Hawks and Doves; this he called an evolutionary stable strategy (ESS). It would describe a population in which any mutation would not knock the population off balance; it would always gravitate back to the ESS. In this case, with the points system I have described, the ESS turns out to be around 58 per cent Hawk and 42 per cent Dove. The mixture is not necessarily the best possible outcome, but it is stable, and that is what counts.

Thus, two very different strategies can live alongside each other; in this case, no other strategy or set of strategies could replace it, no matter how quickly individuals mutate from Dove to Hawk or vice versa. Similarly, Viking society would have suffered greatly had everyone been a berserker, but because only a certain proportion were, it was stable. The individuals in our fictional Hawk–Dove population do not have to stick to the same identity. We can look on the game as played by similar individuals who choose to be Hawk one day and Dove the next. The evolutionary stable strategy also applies to this scenario: if you or I were a player in the game, we could do no better – for our own individual genetic fitness – than to play Hawk 58 per cent of the time and Dove 42 per cent of the time, assuming we do not know who our opponent is in each confrontation.

The Hawk–Dove game has been developed and made more complex so it better resembles a real-life scenario. For instance, it's unlikely that all the players will be equal in their fighting abilities. It is possible to run the game with each individual contest judged on the basis of differing fighting skills, or it's possible to allow some combatants to value territory more highly than others, or we can suppose that in each fight there is already an 'owner' of a piece of territory, who is defending, and the other player is attempting to usurp the owner. This last scenario has the effect of skewing the contest in favour of the incumbent, just as a bird defending its own nest would normally have an advantage.

In the animal world, fighting ability is absolutely central to life in a social group; it determines who wears the trousers. The pecking order in any group of animals will be a fair indicator of who would win in any given confrontation. The construction of a pecking order has the effect of significantly reducing the number of violent confrontations. Among great apes, monkeys and other social mammals such as wolves, body language will signal immediately an individual's 'rank' in the system: dominant animals will have a more erect posture and will tend to be nonchalant and confident in how they carry themselves; subordinates will appear nervous, averting their gazes and lowering their heads. As we know, these outward signs are common in humans. Think of the last time you had an argument or physical confrontation. It might have been a school-yard scuffle, a quarrel over a spilt drink in a bar or a row at work.

Try to remember your tone of voice, your gestures, your posture. Were you playing the Hawk or the Dove?

Game theory is a persuasive method of exploring the 'cold calculus of evolution'. It is based on the idea that evolution is completely dependent on the survival and reproduction of particular genes, and as a model of natural selection it can prove the point that the evolution of behaviour is not obvious, nor is it a foregone conclusion. Game theory shows us that violence was potentially a useful strategy in our ancestral environment of small groups on the savannah. It was the best possible solution for individual success, at least in certain situations, such as defending resources, territory, or a mate.

But what Maynard Smith, and countless other researchers who have delved deeper and deeper into the complexities of game theory, have shown is that more than one type of behaviour can co-exist in any given population. Evolution does not mean we all have to act the same way, nor does it mean that our behaviour is consistent. We can be Hawks one day and Doves the next. What is important is that, in the long run, the mix of strategies is stable.

The killer ape

If we looked hard enough, we'd find that the animal world is full of games, although most are so complex that our model of their behaviour can only be an approximation. So far, though, we have examined only one-on-one conflict. What about organized violence?

Until fairly recently, organized violence within one species of any kind was thought to be the exclusive preserve of human beings – one out of four thousand species of mammals and ten million species of animals. Studies of animal behaviour seemed to reinforce this belief. When Jane Goodall, the acclaimed primatologist, began studying chimpanzees in the Gombe National Park in Tanzania in the mid-1960s, she found a scene of sophisticated tranquillity and co-operation. The chimps were fun-loving and playful; there were battles for supremacy among the males which sometimes turned violent, but rarely did any of the chimps get injured. They would hunt monkeys together, and when one was caught the meat would be shared among friends and allies. There were rivalries and petty jealousies, but the group remained close-knit and seemingly content with their lot.

But a few years later Goodall realized that the original group was beginning to split into two, each with its own territory. As the dominant personalities attached themselves to each group, the rest of the chimps were forced to take sides. Soon after the split came a discovery which shocked all seasoned chimp-watchers – the battered and beaten body of an adult female. The only explanation seemed unthinkable. Had she really been murdered by the rival group?

A short while later, there was another killing, and this time one of Goodall's field assistants was there to witness it. A chimp called Godi, a twenty-one-year-old male, was crouched in a tree, eating. He was alone, unusually, since he normally travelled in a group. Eight chimps from the rival group spotted him from some distance, crept towards the tree and pounced. Godi jumped and ran, but there were too many attackers and he was quickly overwhelmed. One of the chimps pinned him on the ground while the others brought their powerful fists down on him, one even using a stone. After several minutes of the most severe punishment, the chimps ran off, hooting and screaming. Godi's body was bruised and bleeding profusely; he was never seen again, and almost certainly died soon after the attack.

Subsequent murders of chimps were also witnessed and described, and a pattern began to emerge. Chimpanzees from one of the groups would take to prowling the perimeter of the territory, waiting for a solitary target, a chimp from the rival group. If they saw one, the raiding party would fall silent and stalk their enemy like deerhunters tracking a kill. Then they would attack with lethal force, their whoops and screeches sounding like battle cries. By 1977, just a few years later, all the members of the second breakaway group had either been killed or were forced to rejoin the first.

There was extraordinary premeditation involved in this so-called 'lethal raiding'. It seemed that one male chimp would become excited and then a border patrol would form. The group would then leave their base together and set out for the edge of their territory.

Sometimes they would call out, waiting to hear a response from the other group. If they did, they'd often head back to their own range and the rest of their group, but if during their secret foray into the neighbouring territory they encountered a solitary and vulnerable chimp from the other camp, their reaction would be very different. They would halt their calls and start silently stalking their lone target. They would then launch their attack with a sudden and violent ambush – and complete co-operation from all members of the patrol.

At the time, animal behaviourists were stunned to hear of strong and repeated evidence of such violence among the Gombe chimps. Until this point, most people believed that organized killings occurred only as a result of warfare, which had only been seen to take place between humans and had never been observed elsewhere in nature. They wondered if the chimps' behaviour had been brought about by the unnatural circumstances of their interaction with human researchers. Upon first setting up her camp at Gombe, Jane Goodall had decided to supply the chimps with bananas as a means of being able to study them at close quarters. While critics claimed that this provision of food had upset the chimps' normal levels of competition, causing other troops to invade the area, this argument was later dropped when other studies observed 'lethal raiding' in research areas where there was no food supply on offer and human interference was at a minimum.

So why would chimps go on these lethal raids? Probably because it is useful to knock off a member of a neighbouring gang; the reasons boil down to securing territory, food and females. But for each raid there is a substantial risk attached for the border patrols. The victim could fight back and inflict serious damage on members of the raiding party. The chimps take this risk into consideration: only if they outnumber their target and have the element of surprise do they launch the assault. Organized violence takes its toll: among the Gombe chimpanzees lethal raiding accounts for at least 30 per cent of chimpanzee deaths. It seems as if our closest primate cousins do indeed share a natural instinct for violence with us.

The human warrior

Ever since Hobbes there has been little agreement on the question of whether hunter-gatherer cultures were peaceful or warlike. Some anthropologists have taken the view that warfare began in earnest only with the beginnings of civilization. Prior to agriculture, they said, most tribal or hunter-gatherer societies were essentially peaceful, and they pointed to the many examples of non-violent cultures. The !Kung, they claimed, hated violence and believed that anyone who fought was 'stupid'. The Mbuti people, nomadic pygmies from Central Africa, also gave violent people no credit; they never even alluded to it in their story-telling or dancing (Punch and Judy, on the other hand, is a good example of story-telling that *depends* on violence for dramatic effect, and you've only got to look at the box-office receipts of Hollywood movies to see how attractive violence can be to a modern audience).* The warlike cultures of some native Americans were thought to be a fiction used by Western colonialists to justify their own violent conquests. Other anthropologists suggested that colonialism created violence where there was none before; the American Indian Comanche, for instance, were supposed to have been a peace-loving people before their contact with the Europeans. Others pointed out that some peaceful cultures had radically different conceptions of territory and possessions. It was reported that the !Kung had no boundaries or borders to their territory and made no attempt to defend it; Aborigines seemed to have little concept of private property, and their attitudes embodied an ethic of 'share and share alike'.

But a rival anthropological faction has claimed that all hunter-gatherers were warlike, and many still are. This view presumes that war is a 'natural' condition of human societies and that peace is an unusual and temporary state of affairs. Certainly war is extremely contagious; if a neighbouring tribe or clan attack, the only way to

* The amount of violence in the media has led to a new field of research examining whether this is a cause of violence in society or a trigger for underlying violent tendencies. In one study, the American Psychological Association estimated that the average American child sees more than 100,000 acts of violence and 8,000 murders on television.

survive is to defend oneself and probably to launch a counter-attack. Perhaps some societies have been wiped out as a result of being unwilling or unable to engage in violence. One would expect war to spread, like dominoes collapsing one after another.

In reality, though, neither of these extremes can be supported. Anthropologists seem able to find as many peaceful societies as there are warlike ones. Some claim the mix is roughly fifty-fifty, others that a significant majority engage in warfare, no matter how infrequently. What is clear is that humans are capable of both war and peace, although peace is of course far more difficult to maintain as a long-term strategy.

There are some convincing evolutionary explanations of warlike behaviour. In most wars the victors end up with a better chance of passing on their genes than the losers. For example, the practitioners of warfare in many ancient civilizations placed a premium on killing the men and kidnapping the women, especially virgins, who could carry children for the victors. After crushing the Midianites, Moses urged his officers to keep alive only the virgin girls. A somewhat similar tactic had been pursued by the Egyptians, whose Pharaoh ordered every son born to the Hebrews to be thrown into the Nile. And abduction or rape of the losing side's women was often a feature of conflicts. From this perspective, war is a genetic battleground; there is everything to be gained and everything to lose.

Critics of an evolutionary explanation of warlike behaviour point to the enormous role culture plays in the development of warfare. Even chimpanzees are capable of learning behaviour, and lethal raiding could have just 'appeared' one day in the calendar of chimpanzee activities, for reasons we may never be able to fathom. There is no proof that chimpanzee lethal raiding is genetic, nor that it evolved as an adaptive form of behaviour.

But chimpanzees are our closest animal relatives, and we are the only two species to engage in organized violence. I believe *Homo sapiens* has evolved a capability to be violent and that we are able to *choose* warfare as a strategy, depending on the conditions and ecology in which we live. Our capacity for warfare appears to be adaptive, and it seems logical that natural selection has rewarded those who had the instinct to fight or defend when attacked. And once one group of early humans takes up arms, the rest have to follow; they have no choice if they want to live and thrive. 'War is like love,' said Bertolt Brecht, 'it always finds a way.'

Human history has been driven by warfare. Once we settled down to farming and civilization ten thousand years ago, mankind has become extremely good at waging war. We need only to trace the movements of peoples, borders and nations to realize that war is the main process by which the modern world has been constructed.

As technology developed, our ability to conduct large-scale organized violence was refined. We began fighting with our bare hands, then picked up stones, then learned how to sharpen stones and make handaxes. Some fifty thousand years ago we invented spears, then bows and arrows, and more recently catapults, cross-bows, cannons and finally guns. As the scale of war grew, soldiers fell victim to increasingly advanced weapons. Our intellects and superlative problem-solving abilities have been put to work constructing increasingly extraordinary methods of fighting and killing our fellow humans. Hiroshima stands as the appalling climax so far of mankind's lethal technology. Einstein pointed out the scale of destruction made possible by his discoveries: 'I know not with what weapons World War Three will be fought,' he said, 'but World War Four will be fought with sticks and stones.'

In the meantime, though, technology, centralized organization and control serve to take war further and further away from the skirmishes and stone-throwing of the savannah. For the men bedded down in muddy trenches in World War One, the reasons why they were fighting must have seemed very distant and unclear. That is why most large-scale warfare has to be conducted using the well-worn tools of propaganda and coercion in the form of conscription. Soldiers do not stride willingly off to fight for no reason. If the compulsion or the propaganda falters, fighting does not seem so worthwhile; when the Russian tsarist regime collapsed in 1917, for instance, Russian soldiers simply refused to kill any more Germans.

Whether an army is made up of willing fighters, reluctant conscripts or even brainwashed drones, the mechanics of warfare itself depend heavily on co-operation. Without planning, communication and co-ordination an army would cease to be a useful military force. Co-operation is the subject of the next chapter, but we shouldn't forget that co-operation and conflict are two sides of the same coin.

More important, though, than the technicalities of warfare is that co-operation creates a common identity and a sense of safety and fellowship. We constantly make the distinction between 'them' and 'us'; we differentiate ourselves from the group in the next valley or in the next country, or from a group that practises a different religion. Even the passions that arise as a result of supporting your football team are rooted in our instincts for splitting up the world into friend and foe.

The consequences of this ubiquitous human trait are brutal. Think of the Holocaust, the Spanish Inquisition, the horrors and ravages of colonialism; Rwanda, Kosovo and Afghanistan; world wars, civil wars, ethnic conflicts, the Cultural Revolution and Year Zero. Perhaps, then, Hobbes got it all wrong. Modern technology and organization coupled with our ancient human instincts are now allowing mankind to become ever more violent. The State of Nature, in comparison, might well have been an idyll.

chapter**seven**

Co-operation and Altruism

The hermits

For most of us the life of a Maronite hermit would be an unappealing prospect. Rising at the first light of the Syrian dawn, shrugging off aches and pains inflicted by a rough stone floor, the hermits' day began with hours of prayer and contemplation. Then they might have eaten a simple breakfast, perhaps the only meal of the day. Some hermits spent their afternoons copying religious manuscripts. They would have sworn off lovers and companions; their only human contact would have been with the occasional pilgrim who came seeking spiritual guidance or a blessing. It was a life that required the complete renunciation of the material and human world.

Over the years the solitary life has proved strangely popular. The Chinese were encouraged by the *Tao te Ching* to enjoy the pleasures of silence and the delights of nature. Orthodox Russian hermits settled in the vast forest wilderness of Siberia, where some won reputations as oracles and attracted visits from Russian intellectuals and statesmen. The Maronites were not the first, but they may well be the last. Just two Maronite hermits still live as ascetics in the mountains of Lebanon, the bearers of a two-thousand-year-long tradition.

In eighteenth-century England this lifestyle became positively fashionable for a time. Well-to-do members of society could hardly become hermits themselves, of course; after all, they had a stately home to manage and friends to entertain. Instead, they employed hermits to live on their estates, providing amusement for their guests and even an occasional spiritual insight. Enthusiasts wrote articles on how to build a hermitage, from a basic hut intended to reflect the primitive conditions of the Doric order to the more luxurious Chinese model, which is built around a large tree and is fully furnished with a couch and seating area. So-called 'ornamental' hermits were not easy to find, however, and patrons went so far as to advertise in the personal columns of national newspapers. The owner of a country house in Payne's Hill, Surrey, stipulated in his advertisement that his hermit should serve no fewer than seven years in the hermitage, where he would be provided with a Bible, spectacles, food, water and a mat on which to sleep. He was not to cut his hair or his nails. The advertiser was not concerned that a true hermit was unlikely to be reading the newspaper personal columns on a regular basis. The successful applicant lasted only three weeks; he was sacked after being caught in the local pub.

Few of us have the temperament or the stamina for such a lonely life. Like Miranda, Prospero's daughter in *The Tempest*, whose only companions on the island were her father, a sea-monster and an invisible spirit, we would be in desperate need of human contact. 'O brave new world,' she cries when the castaways wash up on shore, 'that hath such people in it!' (''Tis new to thee,' replies her father ironically, knowing how much noble appearances so often belie the base instincts of the human race.)

We are social beings at heart, and we know it. 'Man is born for society,' said Denis Diderot, the Enlightenment philosopher. 'Separate him, isolate him and his ideas will become disjointed, his character will change, a thousand ridiculous emotions will rise in his heart, extravagant thoughts will rise in his spirit like brambles in waste land.' Most of us would agree that a life without friends, family or lovers is unimaginable. Social contact is at the core of our daily lives. The challenge is to explain why.

Selfish gene theory

Some species come together simply to mate; they spend most of their time as lone individuals, feeding, sleeping, sheltering and killing as self-sufficient organisms. On the Pleistocene savannah, humans would have been virtually incapable of living alone, not necessarily because of an emotional need to have people around, rather through a sheer inability to survive. Finding food, battling the elements, warding off predators and bringing up children demanded co-operation. It is likely, therefore, that material needs came first. We must examine whether the mind has adapted to seek co-operative and trusting alliances with people around us. Do we have an instinct for collaboration and sociability? Is not human nature essentially selfish and self-interested?

In 1976, the evolutionary biologist Richard Dawkins published his most famous and brilliant book, *The Selfish Gene*, and the work has often been misunderstood ever since. His metaphor of the selfish gene was founded on William Hamilton's work in the early 1960s which emphasized the role of the individual gene as the focus of evolutionary change. As a reminder, the following is a crude version of his thesis. It begins by stating that any gene that promotes its own survival and replication will spread at the expense of other genes. In other words, genes that do well – genes that spread throughout a population – do so because they are, in a sense, self-interested and further their own chances of replication. Genes, of course, do not have any thoughts, feelings or desires, conscious or unconscious, so we should rephrase this: successful genes operate *as if* they are self-interested entities. This is the first half of the selfish gene theory. The second half goes like this. Organisms are built by genes. Therefore organisms are machines created by genes to enhance the chances of the genes' replication.

This is an extremely powerful idea, one that even Richard Dawkins himself says continually surprises him. But first let me add another caveat. Genes are not the only factor in the development of an organism. Genes are in many respects quite distant from the coal-face of development. They send protein messengers that instruct other proteins to build structures in the body or the brain, and in different organisms the same gene does not necessarily produce exactly the same structure – consider the small physical differences between 'identical' twins. Moreover, the environment can have a considerable effect on how an organism develops.

Broadly speaking, however, the development and final form of an organism are largely dependent on the genes, and they are changed by mutations within the genome. This leads inescapably to Dawkins' conclusion. To understand the implications of his point of view it is useful to imagine the very beginnings of life on Earth. Before life began there was just the primeval soup, and in the soup there was a mixture of molecules, particularly carbon dioxide and methane, both carbon-based, and water from the gases hydrogen and oxygen. From these simple inorganic molecules and atoms it is relatively easy to make amino acids. Such more complex organic molecules can be synthesized in laboratory conditions that mimic the kind of environment we think existed soon after the beginning of time, an atmosphere with regular electrical discharges, high temperatures and ultraviolet light. It is postulated that these amino acids – the building blocks from which all proteins are composed, essential to life on this planet – started to form chains. Similarly, relatively simple carbon-containing compounds such as sugars and purines were also formed. All these substances are the basic components of nucleotides, and hence DNA. Proteins were made which possibly broke up and recombined randomly. Eventually, by chance, there appeared a few molecules that could make copies of themselves. By combining other amino acids as raw material, they happened to be able to copy their own chemical structures. These molecules, which we can call replicators, spread throughout the primeval soup, reproducing and increasing in number.

But some replicators did not last long. The replication process might have been flawed, thus they made copies that were riddled with errors; the errors built up and eventually, in the final generation of copies of copies, the molecules could no longer replicate themselves. Like a lengthy game of Chinese whispers, the last molecule in the line was effectively gibberish. Others might have been able to make exact copies, every single time, but they would never change and never evolve. They would produce molecule after molecule, copy after copy, without ever developing into something more complex, until they ran out of the more simple proteins that provided the raw material for replication. So exact replication did not guarantee success. Once again, they were an evolutionary dead end.

Long-term success needs a happy medium. It needs a molecule that replicates and makes *just enough* errors to evolve and adapt to the conditions in which it lives. If one particular variation of this replicator runs out of raw materials, or maybe gets 'eaten' – used as raw material to feed another kind of replicator – there will be other variations which may be more successful. Just like the bacteria that mutate and develop immunity to antibiotics, these early replicators will develop immunity against being eaten.

Eventually there was one especially prosperous kind of replicator. By chance, the mechanics of this molecule struck a perfect balance in the copying process – not too exact and not too careless. This replicator had enough adaptability to spread, flourish and eat up all the other kinds of replicator while still retaining its own core ability to reproduce itself. We have a name for the modern descendants of this fantastically successful replicator – DNA. Its offspring are contained within every single cell in every single plant and animal on this planet.

Along the long, twisted double helix of the nucleotides of the DNA replicator lie the parcels of coded information we call genes. Over time, genes mutate and change randomly. Most mutations are pointless, or, worse, are damaging to the survival chances of the genes, but some are helpful, and evolution has shown which genes are good and which are bad.

There is no species of free-swimming DNA in the world's oceans. The closest that we know about are simple viruses, strands of replicators enclosed by a protective sheath of protein. The instructions for making the protective jacket are encoded in the DNA and these organisms are the precursors of every other kind of animal. The point is that every organism, including human beings, is a kind of protective jacket for DNA. Our cells, which in turn group together to make tissue, bone, skin, blood and nervous systems, are all integral components of our DNA's extraordinarily complex survival machine. This view is the logical outcome of gene-centred evolution, a bewildering yet unavoidable conclusion. The human body and mind are adaptations 'designed' to further the survival of our DNA replicators.

It has long been said, and rarely meant seriously, that a hen is only an egg's way of making another egg. No-one realized how close this adage was to the truth. Metaphors can go too far; as the geneticist Steve Jones remarked, 'Evolution is to analogy as statues are to bird-shit'. Possibly Richard Dawkins might be criticized for his evocative labelling of genes as 'selfish'. However, the central idea – that genes are intent on their own survival – is extremely apt. The individual organism merely becomes a machine for carrying the genes, and the microbe, plant or animal is not the real focus of natural selection.

Why does Dawkins call our genes selfish? They are selfish because their *raison d'être* is their own survival. The fact that they exist in a current genome means that they have replicated more successfully than their rivals, which are now extinct. Natural selection is all about the spread of successful genes at the expense of not-so-successful genes. These genes do not have to make life pleasant for their or anyone else's survival machines. As long as the genes can replicate, life for its host organism can be miserable, painful, or just plain dull. Genes do not *care*. The male redback spider who allows himself to get eaten during copulation is following genetic instructions that have already proved themselves good at making as many copies of themselves as possible; no matter that their host ends his life as lunch for a hungry mother-to-be.

For Dawkins, the gene is not just the primary unit of selection, it can be seen as being a self-interested entity whose overriding ambition is its own survival. Selfishness is therefore a cardinal quality in a gene in which paradoxically its own survival comes before the survival of the host organism. A case in point is the existence of sterile worker castes among social insects.

Many people assumed that because our genes are selfish, so are human beings in their relationships with one another. This is only partly true. Siblings will fight over a last chocolate biscuit. Lovers will fight over commitment or infidelity. Rivals will fight over sexual partners, sometimes resorting to violence, sometimes using psychological ploys or deceit. We all compete over territory, each of us attempting to gain ground at the expense of the other; neighbours squabble over garden boundaries, modern military states clash over disputed land. We are, in these many respects, solely interested in our own well-being, or the well-being of our group, and this self-interest is backed up by a sophisticated armoury of tools and tactics.

But selfishness is not the whole story. Gene-centred evolution – as described by Hamilton, Williams, Trivers and Dawkins – tells us that co-operation and altruism can often be as beneficial to our genes as can competition. After all, our bodies and our brains do not just blindly follow one genetic blueprint. We are 'designed' to react in different ways to different situations, and, indeed, in different ways to similar situations. On the savannah our genes played a three-million-year-long game of roulette, random mutation after random mutation, most failing to get the winning number, but a few mutations found themselves useful, even essential, and perpetuated the mortal coil of which they were a part.

In our case, these genes were best served by upgrading their early model host, *Australopithecus,* into the sociable, co-operative, group-living *Homo sapiens* we are today.

Kin altruism

The huge dunes of sand that can be seen along the palm-fringed beaches of Central America contain the nurseries of what must be the worst mother in the world. Each nest, dug some two feet or so into the damp sand, is a dark hiding-place for up to two hundred eggs laid by a female leatherback sea turtle. But having used her massive paddle flippers to scoop out this shallow hole which resembles a shallow grave, she does not stay around. In the dark before dawn, she lumbers back into the warm sea. Around sixty days later, her little hatchlings, miniature, soft-shelled turtles each less than three inches in length, break the shell of their eggs and crawl out of their fragile white hermitage. Digging up through the sand, they take their first tottering steps on a journey of perhaps a hundred metres down the beach to the ocean. These vulnerable little babies must find their own food and fend for themselves. Most will be picked off by birds of prey, lizards and crabs before they even reach the water. Once they reach the sea, a barracuda or a dolphin can swallow a baby turtle in one bite. Their chances of surviving to adulthood are poor. Of those two hundred eggs, an average of two from each nest will survive their youth and grow to sexual maturity.*

Human beings take the opposite tack. We have very few children and make an enormous investment – in terms of time, energy and resources, even placing ourselves at risk of injury or death – in each of them to ensure their survival. Our generosity even extends beyond our children to our close relatives, but the closer the blood relation the more we are willing to do for them. This is a direct result of kin selection and an inescapable consequence of the fact that we share the most genes with our children, our parents and our siblings.

We have already seen how social insects provide a vivid example of the degree to which kin selection can affect behaviour. It is so powerful a force, in fact, that worker ants and killer bees will undergo kamikaze missions without a moment's hesitation. We have also seen the terrible consequences of 'anti-kin' selection – male lions killing cubs in their adoptive pride. We, too, are influenced by the mathematics of kin selection, which predict that a parent who shares one half of his or her genes with a child would be willing to die for two children or two siblings (or four grandchildren, or eight cousins). But behaviour does not adhere to such mechanical calculation. Kin selection gives us a predisposition only, one that makes us more likely to place ourselves in danger to save our children. When a mother dives into a surging river to save her infant, she does not consciously or unconsciously make a calculation; she simply dives in (or not, as the case may be).

Family ties account for a part of our social instinct. The nuclear family structure, such as we see in many birds and other mammals, might have marked the beginnings of our sociability. Modern

* These sea turtles are an increasingly endangered species because of the most intelligent predator of all. In a poor country such as Costa Rica where people scratch around to make a living, turtle eggs are much prized as a valuable source of income and this coastline is becoming increasingly depleted.

hunter-gatherers like the Inuit or the Pygmies of the Ituri Forest live in groups tightly bound by family links. Their social organization is fluid. Sometimes they will live in so-called 'minimal bands' of around twenty-five or thirty people containing half a dozen nuclear families, but all are firmly connected by family relationships. When food is scarce these bands may split up and each family travel alone. Equally, a number of minimal bands may gather together in one place if there is plentiful food.

So, does Stone Age family life provide the basis for our modern co-operative instincts?

The Big Mistake hypothesis

John Tooby and Leda Cosmides have suggested that kin selection is the force behind our instincts for large-scale co-operation in the modern world. This idea has been called the Big Mistake hypothesis.

The authors propose that because we evolved in small, kin-based groups on the Pleistocene savannah we have a predilection for co-operating with *all* members of any given group. The given reason is that on the savannah most of the group would have been related by blood, or at least by 'marriage' – that is, sexual or child-raising partnerships. Our minds are adapted for the savannah rather than the enormous and complex societies we live in today, so now we 'mistake' wider society for kin; therefore, co-operation in modern society is a maladaptive by-product of kin selection. In other words, what used to be a useful evolutionary strategy for several million years has now backfired. We cannot completely get out of the habit of co-operating with those around us, even though now the benefits for us are far less, since there is no genetic self-interest in helping strangers who do not share a significant proportion of our genes, unlike our close family. Moreover, strangers do not have a genetic interest in helping us or returning a favour, which means that we could be suckered into helping someone and getting nothing in return. Broadly, therefore, we are all more or less co-operative because we all have the same maladaptive mechanism.

Presumably this is why I and my son Ben shout ourselves hoarse from the season-ticket holders' seats at Arsenal Football Club. If I am strictly honest, genetically speaking I feel I have remarkably little in common with the somewhat foul-mouthed, heavy-smoking, oafish and overweight members of the human race who regularly sit on either side of me at the Highbury ground,* but when Dennis Bergkamp flicks an elegant pass to Thierry Henry in front of the Tottenham Hotspur goalmouth, I simply want to hug those wonderful next-door neighbours.

But of course, for this very reason the Big Mistake hypothesis is not wholly convincing. Our small, intimate bands or clan groups may have been made up of kin, but for any family member more distant than parents, children or siblings our genetic interest drops off extremely rapidly. Cousins share only one eighth of our genes; with our cousin's children we share one sixteenth. This exponential drop-off in genetic interest must mean that our immediate family should, according to kin altruism, benefit from virtually all our altruism or co-operation, even in a relatively small group, and that we have an instinctive mechanism for sorting close kin from distant kin. This is not necessarily the case. If you were brought up in an extended family and not told which of your peers were siblings, which were half-siblings and which were cousins, there is no in-built 'kin detection' system to figure things out. The necessity of incest avoidance, relatively important because of the health implications of inbreeding (and vitally important over several generations), would seem to suggest that we *can* tell kin from non-kin, but studies of children brought up communally on Israeli kibbutzim show that very few members of any given peer group, whether or not they were kin, ended up in a sexual relationship or marriage. It appears possible that familiarity during childhood, rather than biological relatedness, is the key to incest avoidance.

In any case, much of modern behaviour contradicts the Big Mistake hypothesis. In modern society we tend to favour kin over

* For reasons of personal security and Ben's peace of mind, I shall refrain from giving you, gentle reader, my seat number.

non-kin, despite our tendency to co-operate with all or most members of society. We tend to leave more money to biological children than non-biological children; we make enormous sacrifices for close family, even if we have never met them: we hear stories of brothers or sisters separated at birth who, years later, donate a kidney or bone marrow in order to save the other one's life. In my own field of medical work, people go to extraordinary lengths and often crippling expense in attempts to have their own genetic children, rather than adopt. Consciously and unconsciously we all make the distinction, although with the splintering of the nuclear family and greater mobility, biological ties have become somewhat more fragile.

A cursory glance at government appointments in both democratic and despotic regimes around the world reveals that nepotism is still a powerful force. George W. Bush did not get to be President of the United States on the basis of outstanding political talent or a heavyweight intellect. The ruling elite of Iraq is an even better example. Saddam Hussein's younger son, Qusay, is a member of the ruling party's executive and controls the Republican Guard; his elder son, Uday, runs Iraq's most influential newspaper as well as a paramilitary force known as Saddam's Commandos. Either one looks like a possible candidate to succeed his father as president.

Reciprocal altruism

So, kin altruism can take us only so far. Co-operation among animals has evolved not only between non-kin; this spirit of free trade also extends to relationships *between* species. One example is the cleaner-fish, the many species of small fish and shellfish that make their living picking off dangerous bugs and parasites from the skin of larger fish. Some even venture inside their hosts' mouths, cleaning teeth and the insides of cheeks. The larger fish almost never eats the cleaners, even though many of them are small enough to be swallowed in a single, sneaky bite.

In *The Selfish Gene*, Richard Dawkins describes a wonderful example of reciprocity between species. Some species of ant have struck up a fertile relationship with aphids, which are extremely good at sucking the sap out of plants and extracting the nourishment. The aphids need only a small proportion of the nutrients; the rest is ejected from the aphid bottom as sweet, viscous 'honeydew'. Normally, honeydew drips down to earth and goes to waste, but the ants have evolved a capacity to 'milk' the aphids: they stroke the back end of the aphid and catch the honeydew as it flows out (some aphids eject their own body weight in honeydew every hour). The aphids are quite happy with the arrangement. In return, they get protection from predators. A few species of aphid will even allow the ants to take their eggs underground and care for them, bringing them outside to the milking area when they are ready to take their place on the honeydew production line.

Other species find co-operation essential when it comes to outwitting common predators. Zebras are notoriously near-sighted, although to make up for it they have especially keen hearing. They sometimes join together with wildebeest and giraffes, animals that have extremely good eyesight, thus combining their respective talents.

Co-operation on the savannah

For humans, the savannah was not an environment in which one could easily live alone. Food was unlikely to be plentiful. There were fruit, roots, tubers, nuts and berries, but as the forests dried out and receded at the end of the Miocene era and the cold Ice Age air swept down from the desolate north, vegetation became more sparse and increasingly difficult to find. Deer and antelope walked the plains but they were not easy to catch, and the competition was fierce. If a *Homo habilis* was not quickly at the site of a lion kill, hyenas or vultures might strip the corpse clean in minutes.

Early humans who banded together in a group were more likely to survive periods of drought and famine. Members of a large group could cover much more territory, and the chances of bringing home fresh meat and other food were vastly increased. But probably the main reason was protection from predators. In a group, one is less likely to be picked off by a passing sabre-tooth tiger (see box).

In the mid-1980s Craig Reynolds wrote a computer programme that simulated the evolution of flocking and herding in animals. Since then, 'A-Life' or Artificial Life programming has become an extremely useful tool for all kinds of researchers, from evolutionary biologists to engineers trying to perfect the internal combustion engine.

Computer simulations clearly show why flocking is highly likely to evolve, particularly where predators are involved. The reason is as follows. One target animal on its own, say an antelope, is exposed to a sabre-tooth tiger on all sides. If that animal is given a simple rule such as 'steer towards another target animal and maintain a certain distance from it', it will be exposed only on three sides rather than four. The likelihood of the antelope being picked off by the tiger is much reduced. In these simulations we can stipulate that successful tactics will reproduce and spread throughout the population, as they would in natural selection. Rules that tell the target animals to move alongside others will reproduce rapidly, and soon all the animals will be trying to move alongside other animals so that each is surrounded on two sides, then three, then possibly all four. In a well-formed herd only a proportion of the 'flockmates' will be exposed on the fringes; the rest will be completely protected.

Reynolds found that in the programme only four simple rules were necessary to mimic the actions of a moving herd:

1. Separation: steer to avoid crowding local flockmates.
2. Alignment: steer towards the average heading of local flockmates.
3. Cohesion: steer to move towards the average position of local flockmates.
4. Avoidance: steer to avoid any physical obstacles.

If each individual follows these rules, the behaviour of 'A-Life' herds can appear uncannily realistic. Animals on the fringes of the herd will constantly move in towards the centre, so there is a continuous throughput of animals from the centre to the fringes. The model mimics the movements of fish in shoals and flocks of birds as well as land animals. It is also similar to the defensive strategy used by the Merchant Navy during the Second World War. Although the convoy system resulted in many losses, the numbers would have been greater, with or without escort cover, had there been no such system.

But flocking or herding is not necessarily an advantage for the group as a whole. The fact that the animals are in a herd does not mean that fewer animals will be eaten – a tiger will not really care that he has access only to animals on the fringes, and he can only eat so many antelopes. The point is, however, that within the group one is much more likely to get eaten if one's herding instinct is missing. This is why the adaptation is inevitable if it takes hold in a population. Having said that, there may be group advantages. A group of animals may, for example, present a more threatening target to the predator; they also have more eyes and ears to monitor potential dangers. But these advantages are by-products of the benefits to the individual.

Delayed altruism

The practical necessity for co-operation, then, is obvious. Equally, it is possible to see why even a simple computer model can explain why we could have come to live in groups rather than as solitary individuals. But unlike the cleaner fish who co-operate with their hosts, our systems of sharing can get rather complex, especially where there is a time lag between the exchange of favours. Say you agree to pay for my dinner on the basis that next month, when you are broke, I will return the favour. The pooling of resources in this way is mutually beneficial, but it does require a certain level of sophistication in the social group; namely, it requires trust, and the ability to remember the actions of different individuals. There is no point giving favours if one cannot be sure of a favour in return from the same individual.

The classic example of delayed altruism in animals is that of vampire bats. Vampire bats, as everyone knows, feed on blood, especially the blood of cattle and sleeping pigs. This method of feeding is not especially reliable because cows will not stand docile and allow the bats to feed at will. Young vampire bats find feeding particularly difficult; one night in three they do not manage to feed at all, and they have a very rapid metabolism. This is a life-threatening problem, because vampire bats cannot go for more than two or three days without food. They become weak extremely rapidly and die of malnutrition – they literally drop dead. But they have solved these problems by practising delayed reciprocal altruism. If one bat is unsuccessful in finding blood, another will regurgitate some of its own meal for the first bat to drink. (Bats drink more than they need to when they are successful, thus having a surplus to give away when they return.) But vampire bats will not regurgitate blood for anyone. They will give blood to other bats with whom they spend time, and who have done them the same favour in the past. These reciprocal relationships stand them in good stead, and they depend on a sophisticated ability firstly to recognize individuals and secondly to remember whether or not they are reciprocators.

Early humans would have greatly benefited from an extremely sophisticated ability to work together, an understanding of one another's point of view, links of trust and reciprocity, and an ability to spot individuals who are not pulling their weight. But it would be foolish to create a rose-tinted view of Stone Age life. By no means all social relationships are co-operative. Some are founded on power, and the pursuit of power pervades most social groupings in the animal world. The fact is that the most physically powerful males or females tend to get what they want by virtue of their size, and the cleverest get what they want through an ability to manipulate others. The proverbial eight-hundred-pound gorilla with a good knowledge of Niccolò Machiavelli would be an unstoppable force, and a natural dictator. Psychological manipulation combined with physical strength is a dangerous and potent mix.

We have seen in the two previous chapters how competition is deeply ingrained in human interaction and how it often leads to violence. Nonetheless, even the most hierarchical human society needs the co-operation of its members – co-operation which is not solely founded on the threat of force. In the uncertain world of the savannah, we needed to get along, and the smaller the group the more reliant any natural leader is on the goodwill and co-operation of others.

Knowing your group

Group living takes brainpower. You need to remember who everyone is, to begin with. Then you must learn the relationships between individuals: who is so-and-so's daughter, are those two having sex, is that guy getting friendly with *him*? And then there is the need to communicate, to forge links of trust and reciprocity, and to remember where you stand in each relationship. Dominance hierarchies in primates show clearly that each individual monkey has to know the position of every other monkey in the pecking order. If they do not, they could get themselves into serious trouble.

Robin Dunbar has proposed that the size of the primate brain is in proportion to its ability to handle a certain size of group. For a number of different primate species, including apes and Old and New World monkeys, he has plotted the relative size of the neocortex against the average size of their social group, which varies between four or five and several dozen. There is a linear relationship between these two variables, and by comparing the size of the human neocortex Dunbar plotted a point on the graph that predicted the natural human group size as 125 to 150.

Each of us has a network of people with whom we are in regular or occasional contact. It will be made up of friends, family members and colleagues, and possibly *their* friends, family members and colleagues. All these are people with whom one could strike up a conversation easily (at least, one would not require any 'getting to know you' introductions), and possibly of whom one may ask a favour. Dunbar suggests that the mean size of this social network is, as his graph predicts, around 125 to 150. Approximations to these figures crop up in many studies of group size among all kinds of different societies, for example Middle Eastern Neolithic villages, the Hutterites of Canada, East Tennessee farming communities, Amish parishes and modern hunter-gatherer clans. Our circle of community, whether we are urban socialites or rural recluses, reflects the limits of our cognitive capacity.

Risk and trust

Yet in modern nation-states, co-operation reaches far beyond this natural group size. Our lives are intertwined like never before, and social, professional and family life is often extremely dependent on massive networks of people acting in concert.

Take one average day for a fairly typical citizen of a nation in the industrialized West. She is driving a car in a built-up area where she accelerates across green lights certain in the knowledge that other drivers (whom she has never met) will have stopped on a red. Her car has been built in Japan or Germany and she places her life in the hands of the engineers (whom she will never have met) who designed it as well as a few local mechanics. Her child is en route to a school run by several dozen local teachers and administrators teaching a national curriculum designed by experts and politicians (for whom she probably did not vote). From there, she drives to an office where she will speak to various people around the world, none of whom she has ever met, to organize sales of goods made in the Far East using currency that is guaranteed by a central bank and shunted around electronically by a vast and complex network of clearing banks. Her morning cup of coffee began life as beans grown by Costa Rican farmers, picked, roasted, packed, shipped and sold by a hundred different people (whom she is never likely to meet) employed by several commercial organizations around the world. Having left her baby with a babysitter she heard would be reliable, she may go tonight to see a Hollywood film made by a collection of strangers (one or two of whom she would love to meet if she had the chance). At the cinema, if she is extremely unlucky, she may trip and twist her ankle when the (Korean-made) battery in the usher's torch fails; this may entail waiting in Casualty with thirty other perfect strangers in order to see a doctor, radiographer and nurses whom she will, all being well, never meet again.

The evening news does not report everyday, successful co-operation – 'Shipment of coffee arrives safely from Costa Rica'; 'Teachers arrive at school in time to teach children'. The logistics and technicalities of the modern world are taken for granted. Instead, we hear about it when it goes wrong: 'Civil war in Costa Rica hits coffee supply'; 'Teachers to vote on three-day strike'.

Virtually every aspect of our lives depends wholly on co-operation and trust, and on taking risks. We trust our insurance company to pay us when our house burns down. We believe in religions that are defined by ancient systems of traditions, laws and stories passed down the generations by vast numbers of followers. With barely a second thought we risk flying in jumbo jets, even though we know nothing of aerodynamics and have never met the pilot. We trust that our country has an army that will defend us if we are invaded, even though we ourselves have no knowledge of military strategy or technology. In return, our governments are given the right to imprison us if we infringe the law. Even the Maronite hermits are bound up in intricate webs of collaboration, trust and regulation.

Free-rider problem

Large-scale co-operation is not always easy to achieve. The easiest kind of co-operation is the convention, an arbitrary rule or law that exists to make everyone's life easier; for example, we drive on the left, instead of choosing randomly between left and right each time we leave the house. (It would be interesting to see how long it would take a country to collectively 'decide' to drive on the left or right if there were no convention or law already in place.)

There are a thousand and one other conventions that are not so obvious and not so clearly stated. Language is the first and most important. Language is an astounding feat of co-operation; it only works if everyone involved follows the same rules.

Other kinds of co-operation are, in a sense, more difficult. We vote, elect governments, contribute to the health services or education. These activities appear co-operative and seem designed to contribute towards a common good, but game theory predicts a problem. Say we have a group of Co-operators, people who consistently contribute towards a common project – lifting heavy rocks from one place to another, for example. They may be building

Stonehenge, or a public toilet, but the end result is something that will make everybody's quality of life a little better, and the benefits will easily outweigh initial costs in time and energy. This particular toilet, however, is an impressive structure made of granite blocks; they are heavy, and it takes at least a dozen people to lift a single block.

Imagine that one of the people in the group decides he has done enough work for the day. He pretends to help lift the block, but in fact he is not carrying any weight. Now, the difference this makes to the other eleven people is minimal – not negligible, for the weight they carry increases by one eleventh, but the difference is small enough that they do not really notice. For our Slacker, however, the difference is substantial: he or she saves a great deal of energy and avoids the risk of being injured. From a self-interested perspective, the decision is perfectly rational. Assuming no-one realizes what he is doing, the costs outweigh the benefits.

What if everybody decided to become a Slacker rather than a Co-operator? The stones would not even get off the ground and the project would be dead in the water. Yossarian, the self-confessed coward of *Catch-22*, is aware of the elegance of this argument when he pleads with Major Major not to fly any more bombing missions. But suppose everybody on their side thought like that, asks the Major. 'Then', says Yossarian, 'I'd certainly be a damned fool to feel any other way.' Yossarian is right. If everyone decides to be a Slacker, you'd be a fool not to join them.

The same problem arises for health service taxation, army conscription, mass immunization and any other public good that depends on the vast majority of people taking part. With a large number of people who contribute, the rational thing to do, assuming there is a large enough benefit in doing so, is to become a Slacker. In the long term this is not good. If everyone does the same we would have no public goods – no health service, no army, no public toilets. Still, because we have no control over other people's contributions, the rational course of action is to be a Slacker. Better a Slacker than a subsidizer of other people.

If we were all rational and self-interested there would be no public education, health service, army, or indeed any government at all.

And the argument extends to natural resources such as fish stocks in the North Sea. If each fisherman is left to make his own decision about whether or not to reduce his catch, the rational thing to do is to continue fishing at the current level. The inevitable result is the depletion or elimination of fish stocks.

Slacking on the savannah

Instead of thinking about an individual and his or her day-to-day behaviour, we should return to the point of view of long-term evolutionary strategy. Say, for argument's sake, there is a single gene for co-operation.* Those who have the co-operation gene may be inclined to join groups. Group life is less dangerous than a solitary life; people within it live longer, eat better and have more children. But what if there is a mutation and one day a Co-operator turns into a Slacker?

Slackers, or free-riders, get something for nothing. They get all the benefits of group life without any of the cost – therefore they will, on average, be more successful than the Co-operators. The genes that predispose them to slacking will do well and spread throughout the group. So, in the language of game theory, a group of Co-operators is not an evolutionary stable strategy; it can be invaded by a Slacker, and then Slackers can take over the group. Pretty soon, the rocks are not getting lifted and the granite toilet is not getting built. Not only that, but without Co-operators the group falls apart. All the benefits of group living are lost.

In the language of game theory, the building of the public toilet is a 'non-zero sum' game. A zero sum game is a game in which one player benefits at the expense of the other. Consider two people bargaining over the price of a car. If the buyer negotiates a discount of a thousand pounds, that benefit is matched exactly by the cost suffered by the seller. But the building of our granite toilet (assuming

* It's more likely that there is a network of genes that affects the development of particular neural structures, which in turn predisposes us to co-operate or not. However, we do not know which genes or which neural structures are involved.

it is eventually put to good use) is a non-zero sum game. Each player has to pay out, in time and energy, in order to lift the stones, but the reward of the finished building is worth more than this initial cost. Games that involve co-operation are nearly always non-zero sum games; the final outcome will generally outweigh the sum of the initial costs. We complain about taxation, but most of us would agree that the benefits of schools, roads, police forces and hospitals outweigh the costs of our own contribution.

Prisoner's Dilemma

There is a simple game that can be used to examine the evolution of human co-operation. It is called the Prisoner's Dilemma, and like taxation or building a toilet, the game is non-zero sum.

Imagine a robbery has been jointly committed by two people, Player A and Player B, and immediately after the crime they are caught and brought in for questioning. They are held in separate cells, and the police attempt to get each of the prisoners to implicate the other. For the purposes of the game, imagine there is no prior agreement between A and B; there is no plan to communicate during the game and there is no loyalty between the two prisoners or favours owed. Honour among thieves plays no role in the Prisoner's Dilemma. The aim for each player is simply to act in his or her own best interests.

Imagine you are Player A. The detective tells you that if you implicate Player B, and if B stays silent, not only will you be set free, you will also receive a reward. Player B would, in this case, be given ten years in prison. The reverse also applies: you will get ten years in prison if Player B implicates you and you stay silent. Secondly, if both you and Player B implicate each other then each of you will be given five years in prison. Thirdly, if neither of you talks – in other words, if you co-operate with each other – the police will have no choice but to set you both free.

The game's options are displayed in the diagram overleaf. Player A chooses a row and Player B chooses a column, with the payoffs for each player as shown.

	PLAYER B		
PLAYER A		**Implicate partner**	**Stay silent**
	Implicate partner	A: 5 years B: 5 years	A: Set free & reward B: 10 years
	Stay silent	A: 10 years B: Set free & reward	A: Set free B: Set free

The best overall outcome, viewing the situation from the outside, is for both players to stay silent – in other words, to co-operate with each other. But what is the rational, self-interested thing to do? If Player B implicates you, the best thing for you to do is to implicate him, otherwise you could end up in prison for ten years rather than five; if Player B stays silent, once again the rational thing to do is to implicate him: you will not only stay out of prison, you will also get the reward.

So whatever Player B chooses to do, the best course of action is to implicate him, or defect. Although staying silent (choosing to co-operate) would be better all round, rational self-interest tells you to defect. This is why the Prisoner's Dilemma is so called. Self-interested individuals do not necessarily choose the best overall outcome. Both players should choose to defect, implicate each other, and both will get five years in prison. Co-operation will be undermined by this tendency to defect, because defection pays.

The Prisoner's Dilemma teaches us a lesson that can be applied to many non-zero sum games. If rational self-interest is the rule, then taxes may not get paid, the stones will not be lifted and the public toilet does not get built.

Adam Smith, the father of free-market liberalism, thought the interests of a group of people were best served by everyone acting to further his or her own self-interest. Applied to economics, this meant that each competitor in a marketplace should try to maximize his or her own profits, thus the overall wealth of the nation would be maximized, guided by what he called an 'invisible hand'. The Prisoner's Dilemma proves that this need not be the case. There are many situations in economics that show that self-interest among all the agents concerned can produce the worst outcome overall, from tax dodging to price-fixing cartels.

One may ask the question of why we are assuming the players in the Prisoner's Dilemma are guided solely by their own self-interest.* The reason is that as a result of investigating the evolution of human co-operation we can rely on the fact that genes are 'selfish'; they are successful only if they act in their own self-interest. Ignoring kin selection, and assuming that we wish to find out if co-operation can evolve between strangers, the assumption of genetic self-interest is extremely important. Long-term strategies to co-operate or defect will only be successful if the strategy is in the best interests of the individual.

But for a particular strategy to spread throughout a population, two conditions need to be met. The first is that the strategy must, over the long term, benefit each individual who is using it. The second is that the policy must be stable within the population – in other words, it needs to be an evolutionary stable strategy and must not be susceptible to any other, more successful strategy.

Iterated Prisoner's Dilemma

A single game of Prisoner's Dilemma can take us only so far. To find out how co-operation could have evolved we need to play many

* I went to school at a time when caning was popular among masters but not among pupils. I remember playing 'Prisoner's Dilemma' for real and in terror. Having broken a window, the choice was between a public, humiliating caning and one hour of detention for the whole class. How I solved the dilemma is a private matter.

games, one after the other, using different strategies and examining how those strategies fare in the population. Do they spread through the population, like a successful genetic mutation, or do they die out? Evidently, a strategy that directs the player always to co-operate will not be successful; if the other player decides always to defect, the latter will win every time.

Robert Axelrod, a political scientist interested in the possibilities of game theory, realized that the only way to find out would be to pit different strategies against one another in multiple games of Prisoner's Dilemma (called iterated Prisoner's Dilemma). He organized a tournament, inviting strategies from game theorists, economists, mathematicians and computer scientists. Fourteen entries were submitted. Axelrod turned the game into a simple computer programme and fed each strategy into the system. Each strategy played the game two hundred times in a row against all the other strategies in turn.

Axelrod summarized his conclusions in his seminal book *The Evolution of Co-operation*, which begins by asking the following question: 'Under what conditions will co-operation emerge in a world of egoists without central authority?' Axelrod's 'world of egoists' sounds like Hobbes' State of Nature. Instead of life being brutish and short, however, the Prisoner's Dilemma shows that co-operation can evolve, even if everyone is looking after his or her own interests.

Axelrod's programme gave points for each round of the game. The equivalent to zero prison time *plus* the reward would translate to the maximum five points. Mutual co-operation (zero prison time) would reward the players with three points. Mutual defection (or five years in prison) scores just one point, and unilateral co-operation (or ten years' prison time) scores zero points* (this last score is also called the 'sucker's payoff'). The points were totalled for each strategy over the entire run of games.

Some of the strategies were enormously complex, with all kinds of rules that depended on the behaviour of the opposing player in previous rounds. Others repeated a pattern, switching between

* I am unclear as to whether Axelrod's point system gives appropriate weighting to each possible outcome.

co-operation and defection at certain intervals. One of the strategies was purely random, choosing to co-operate or defect on a whim. But it turned out that the simplest strategy was, over two hundred games, the highest scorer. It was called Tit for Tat, and it beat the rest hands down.

There were just two rules for Tit for Tat: always begin the game by defecting, and from that point on always copy your opponent's previous move. Therefore Tit for Tat tended to co-operate, but if the opposing player defected once, it too would subsequently defect, but only for one round; afterwards, it would immediately revert to being co-operative. The tactic assumes co-operation in others but immediately punishes defection, and the punishment is in proportion to the crime. Hence the name, Tit for Tat. Tit for Tat is nice, retaliatory and forgiving: it is never the first to defect; it retaliates only after the opposing player has defected; and it forgives the player after carrying out just one act of retaliation.

Next, Axelrod played out the contest as though it were an evolutionary process, with the highest-scoring strategies of each round reproducing and spreading their 'genes' through the population of players. Tit for Tat not only ended up as the winner on points, it also spread throughout the entire population at the expense of all other strategies. Niceness, forgiveness and immediate retaliation seemed to be winning qualities.

Tit for Tat is not, however, an evolutionary stable strategy. In some ways it is a little like pyramid selling. It is true that no other strategy could infiltrate and spread if the whole population were playing Tit for Tat (if someone decided to play 'always defect', for example, they would quickly lose out, usually scoring just one point per round instead of the average three for Tit for Tat). But in a population that plays 'always co-operate', Tit for Tat will not reproduce; there is never an opportunity to retaliate, so to all intents and purposes it acts exactly the same as 'always co-operate'. Axelrod himself realized that there is no one strategy that will become an evolutionary stable strategy in all situations, because it will change subtly depending on the strategy of one's opponent.

Prisoner's Dilemma on the savannah and real life

The Prisoner's Dilemma is an extremely simplified model of how co-operation can evolve. Real life throws up some complications. In Axelrod's tournament the victory of Tit for Tat depended on a series of games involving the same two players. It will not work if each strategy plays a single game of Prisoner's Dilemma, because 'always defect' would win. In a population of real human beings this means that for each game there has to be a strong likelihood of the two individuals meeting again, otherwise the strategies cannot be tested. This is extremely likely, given that early humans probably lived in smallish groups of no more than several dozen others.

Situations that have the same logical structure, and the same structure of payoffs, exist in real life. Hunting on the savannah is a perfect example of a game of Prisoner's Dilemma. It is analogous to the predicament of the vampire bats, in that killing an animal may depend on luck, with only a proportion of the group being successful on any given day. If an antelope or deer is killed, it will provide more meat than one person or even an entire family can eat, before it goes rotten. Sharing meat, then, is a sensible strategy, so the luck is spread around the group; no-one goes hungry, no meat is wasted. Co-operation means that if you make the kill your portions will not be as large on that day as if you had the entire animal to yourself, but this cost is outweighed by the benefit of eating when you have had no luck in the hunting grounds.

For modern *Homo sapiens* there are countless opportunities to co-operate, defect or play Tit for Tat: the decision to uphold or break a nuclear arms treaty; cartels or co-operation between companies selling the same products; sharing household chores with your spouse. If co-operation is indeed part of our evolutionary psychological make-up, if it is hard-wired into our brains, then we can suppose that natural selection has worked in a similar fashion to Axelrod's game. Each generation inherits a tendency to co-operate, to defect, or to pursue a mixed strategy. They do not need to make a rational decision whenever they come up against a Prisoner's Dilemma-like situation, they need only act on their instincts. If the strategy is successful, it will increase the individual's evolutionary fitness and enable the strategy to be passed on more successfully to the next generation.

Pre-conditions for the evolution of co-operation

Game theorists are not quite as impressed by Axelrod's work as are some evolutionary biologists. Ken Binmore, a game theorist at University College, London, has pointed out that some credit Axelrod with outlandish achievements, such as, for example, explaining the evolution of all human and animal co-operation; Axelrod's critics are quick to counter that the iterated Prisoner's Dilemma is a special case that involves just two players and only begins to scratch the surface of the complexities of co-operation. However, Binmore rightly recognizes that Axelrod made the connection between game theory and evolution explicit by organizing the tournament as a sequence of steps with the winning strategies 'reproducing' more prolifically than the others. This, he says, was a significant advance.

But game theorists, with their increasingly complex games of Prisoner's Dilemma, have questioned the conclusion that Tit for Tat is always the most successful strategy. A nasty version of Tit for Tat, called Tat for Tit, has been shown to do surprisingly well. Tat for Tit starts off by defecting, and therefore assumes it can exploit its opponent. If it is matched by defection it will try to co-operate, until its opponent once again defects, at which time it will switch back. The final mix at the end of the game depends on the initial conditions, and on how many 'suckers' and 'defectors' were in the original population.

And when the game expands to include more than two people, 'nice but retaliatory' strategies such as Tit for Tat find themselves in difficulty. Here game theory does reflect our day-to-day experience. We are especially good at maintaining a reciprocal relationship with one other person – a spouse, a friend, a client – but in a large group co-operation can become difficult. Groups are much more prone to defectors and free-riders; it is difficult to come up with strategies that strike the right balance between punishing these defectors and maintaining a good level of co-operation.

Cheat detection

We know from our own daily lives that co-operation is not automatic and universal. Our instincts may predispose to co-operate, in certain situations, but we have to realize that there will be many cases of real-life defection.

John Tooby and Leda Cosmides have developed a theory that we have an adaptive mechanism that allows us to tell whether people are cheating. It is an attractive idea. The strategy of Tit for Tat, for example, can only persist in the population if we know who defects, or cheats, and therefore we know against whom to retaliate. Which person is refusing to lift the granite stone; who is avoiding paying their taxes? We need to know this information if co-operation is to be evolutionarily stable and survive as a form of human behaviour. As we know, testing any evolutionary hypothesis is not easy. Evolutionary psychology has no shortage of theories, but successful experiments are harder to come by. Tooby's and Cosmides' experiment is, however, ingenious (although it goes without saying that not everyone is convinced!).

In 1966, Peter Wason invented a logic puzzle called the Wason Selection Task. It was originally conceived as a means to investigate human reasoning, and it has proved to be an extremely useful tool. In one test, four cards, each with a letter on one side and a number on the other, are laid out in front of the subject. Two of the cards show letters on the front and two show numbers.

The subject is told to test the following rule: 'If a card has a D on one side, then it has a 3 on the other.' Which cards does he or she need to turn over to test whether this rule is true? If you do get it wrong, you will be in the majority; only around 10 per cent of people presented with this task are successful.

But now comes the ingenious part. Tooby and Cosmides conducted a test with the following four cards:

DRINKING BEER	DRINKING COKE	25 YEARS OLD	16 YEARS OLD

The scenario is as follows. You work in a US bar, and you need to check whether there are any under-age drinkers in the bar. Each card represents one person in the bar. On one side of the card is the person's age, on the other side is their drink. Which of the cards would you have to turn over in order to check that nobody under the age of twenty-one is drinking alcohol? (The answers are at the foot of page 276.)

Close to 75 per cent of people tested in the 'bar' scenario choose the correct cards, yet the second test has exactly the same logical form as the first; it is simply couched in such a way as to test our ability to spot 'cheats' in a social situation. Under-age drinkers would be 'cheating', and our success in the second test, say Tooby and Cosmides, means that we are particularly attuned to solving logical problems which involve detecting these cheats.

The first, and obvious, criticism of their interpretation is that we are bound to be better at solving any logical problem described in terms of a practical, and possibly familiar, situation rather than in its coldly logical form. But the success rate of the first test is found to be the same even if 'real-life' situations are used. I will give an example of the kind of 'real-life' scenario that would still present difficulties. The rule to test could be 'if someone owns a Porsche, they also listen to Bryan Adams'. Each card shows a make of car on one side, and a singer on the other; the four cards would show Porsche, Skoda, Bryan Adams and Britney Spears. Once again, the first and last cards have to be turned over to test the hypothesis. From similar tests we know that the success rate is likely to be as low as 25 per cent.

The interpretation of these results is not universally agreed upon. Tooby and Cosmides are emphatic in their belief that we have some kind of specialized 'cheat detection' module in our brains that has evolved to cope with these kinds of situations; others insist that the results are a function of our life experience, or a more general cognitive ability to solve some logic problems and not others. Whether or not we have a cheat detection 'module', however, the research does suggest that for whatever reason we are especially good at spotting free-riders.

Joel Winston,* a Ph.D. student, and his colleagues working with Ray Dolan's group at the Institute of Neurology in London, have recently published an interesting finding. They point out that a key aspect of human social interaction is the appraisal of the faces of the people we meet, and whether we see in those faces signs of trustworthiness. So they conducted an experiment. Placing various volunteers in an MRI machine, they attempted to see what happened in the brain when these subjects looked at another person's face and tried to decide whether or not they could trust them. The volunteers were asked to look at 120 portrait photographs of school and undergraduate students, all taken in a similar way to avoid bias. 'Happy' faces were generally regarded as more trustworthy than those that appeared 'Neutral', 'Sad' or 'Angry'. And in the brain scans, our old friend the amygdala and its connections with part of the temporal cortex were shown to be the active parts of the brain when making these judgements. Moreover, it seemed that the amygdala response was automatic and instinctive, while the cortex was activated only when the subjects were asked to make a judgement. These mechanisms presumably have evolved because human survival depended to some considerable extent on our ability to assess other people and then make accurate social judgements.

Altruistic punishment

If we do have an ability to spot cheats – whether it is someone who is taking the lion's share of an ibex kill, or a queue-jumper in a

* I am not sure whether I should declare a personal interest here, but he is an illustrious relative.

The answer in both cases is that the first and last cards have to be turned over.

cafeteria lunch queue – there has to be some punishment of the free-rider if our cheat-detection module is to be of value. A Swiss economist has shown how we are willing to punish those who appear to be getting a free ride, even if the punishment is costly for the punisher.

Ernst Fehr, of the University of Zurich, gathered together several groups of four students. Each student in a group was given a certain amount of money, say twenty Swiss francs. They were told that they could invest in a group 'project', and that for every franc that was invested, the group as a whole would be given 1.6 francs back, which would be split equally between all four players. So, if each member invested ten francs, each would get sixteen francs back, but if only one of the group invested ten francs, they would get back sixteen francs between them, or four each. The investments were made simultaneously and anonymously.

As in the Prisoner's Dilemma, co-operation pays off, but the rational course of action for each individual game is not to invest. Best to be a free-rider, to sit back and collect free money other people have invested. But if everyone did that, there would be no money invested and no money won.

Fehr found that investments started off at around ten francs on average. When the players realized that some members of the group were free-riding, the investments dropped off to an average of four francs by the sixth game. For the second round Fehr introduced another rule. The investments were no longer anonymous, and one member of the group could choose to 'punish' another for free-riding. This punishment cost the free-rider three francs. Crucially, however, the punishment also cost the punisher one franc. The cost turns the punishment into an altruistic act, because only one out of a possible three players is choosing to accept this cost, and the groups were shuffled around so that each player could never encounter the free-rider again.

The punishment system worked. The level of investments went up, and they stayed up, increasing to over sixteen francs. Why would we inflict damage on ourselves to punish a free-rider? The answer, says Fehr, is anger. He found that anger was the overriding emotion of players who decided to punish the free-riders, and it outweighed the monetary cost.

Everyday life throws up many opportunities for punishing free-riders, whether they are individuals, companies or institutions. It may take the form of a letter to a newspaper, or taking part in a demonstration or political campaign. It may simply mean a shouting match with someone who jumps a queue at the checkout tills. Often the costs involved are greater than one's own personal rewards, but as a whole we all benefit greatly. The members of modern hunter-gatherer societies are extremely keen on using social sanctions against cheaters or free-riders. Among the Netsilik, an Inuit tribe on the shores of the Northwest Passage in the Canadian Arctic, all able-bodied men are expected to contribute to the seal hunt. If any individual does not contribute he is criticized, and if his behaviour is especially bad he is ostracized by the entire group. The !Kung San have a phrase: if someone is hoarding food or other goods for themselves, they are said to be 'guarding it like a hyena'. The punishment is for the hoarder to 'give until it hurts'– in other words, to give away the food until he or she is cleaned out. The co-operation and communal system is fenced around by persuasion, insults and painfully severe sanctions.

An old Muslim proverb warns,'Trust in Allah, but tie up your camel.' Our instincts for trust and co-operation are safeguarded by the mechanisms of cheat detection, social norms and punishment. Our camels are well and truly tied up.

Learning to co-operate (rhesus monkeys)

I would like to sound a warning bell. Our capacity for co-operation and sociability is not necessarily deeply embedded in our genes to the extent that we cannot change our behaviour. Let me use an example that shows the potential for the mutability of primate behaviour. The primatologist Frans de Waal, a professor at Emory University who has spent his life researching ape psychology, has conducted a remarkable study on peacemaking and reconciliation among primates. He has described how, in many species of ape, after two individuals have had a fight, they will reconcile with some friendly gestures or mutual grooming; it is an essential form of co-operation that can resolve potentially violent situations. Often a third ape will encourage the two combatants to get together and sort out their differences. De Waal has even seen female chimps approach angry males who are holding sticks or stones and prise the weapon out of their hand. (Other cautious researchers in the field were unhappy about de Waal's tendency to attribute human-like qualities to primate behaviour. They did not even like the word 'reconciliation'; one suggested that he call it 'first post-aggression contact'. Whatever one calls this behaviour, though, its function is clear.)

De Waal set out to investigate whether monkeys could *learn* to be peaceable with an intriguing experiment involving two species of monkey. The first species, rhesus monkeys, have rigid hierarchies, tend to fight a lot, and show greater signs of animosity towards their rivals within the group. The second, stump-tailed monkeys, are more tranquil, and are reconciled after a fight much more frequently – three times more than their rhesus cousins. De Waal wanted to find out if their characteristic personalities would hold firm if they lived in close proximity, so he set up a mixed colony (the rhesus monkeys were slightly younger and smaller than the stump-tailed monkeys to offset the danger of their physically dominating the latter). At first there was not a great deal of mixing; the two species kept themselves to themselves. The rhesus monkeys tried to make intimidating gestures towards their neighbours, but the stump-tailed monkeys did not rise to the bait. Then curiosity got the better of them, and there was a little tentative contact and mutual grooming. The stump-tailed monkeys were especially intrigued by the long rhesus tail. It was not long before both species were playing, grooming and fighting together in one large heap.

They were friendly and playful, but had their behavioural patterns actually changed? Had either group influenced the social mores of the other? Had the idiosyncrasies of the rhesus rubbed off on the stump-tails, or vice versa?

De Waal counted the number of times each species reconciled after a conflict or spat. His conclusion was extremely interesting. The rhesus monkeys had adopted the stump-tailed tendency to make peace, but they did not adopt any other stump-tailed characteristics, such as calls or gestures. And when the two species were separated again, the rhesus monkeys remained that way.

It is rather interesting to consider (perhaps we shall never know) why, in the wild, rhesus monkeys have remained stubbornly bellicose towards their rivals and stump-tails have found a way to smooth the experience of social living. The experiment showed quite clearly their capacity to change their behaviour through learning, and it would be interesting to see whether the rhesus monkeys transmit this 'new' behaviour to their offspring.

I suspect that Frans de Waal was secretly pleased that the stump-tails influenced the rhesus monkeys and not the other way around. It would have been faintly depressing had the stump-tails started to stomp around, pick fights and bear grudges for the first time in their lives. Perhaps co-operation and peaceability are easier to learn; and certainly the rhesus monkeys always had the capacity to act in this way. But it is difficult to judge what the implications are for human society. Our history tells us that war can spread as easily as peace. Both conditions seem to be volatile and impermanent.

Conclusions

Despite the ease with which the rhesus monkeys transformed their reconciliation habits, I think their behaviour changed in degree, not in kind. We know that the intensity of trust and co-operation within human alliances can differ greatly from culture to culture. Perhaps some of us play Tit for Tat and others Tat for Tit; we have also developed the capacity to change strategy depending on our circumstances. No doubt our strategies are influenced heavily by cultural traditions, but co-operation is an extremely strong contender for a place in our instinctual armoury. We are born to live our lives in collaboration with others, and we are born with the ability to realize when others are failing to play their allotted role.

The fact that we have selfish genes does not translate into a Hobbesian maelstrom of violence and rivalry, of every man or woman for themselves. Richard Dawkins wrote, 'be warned that if you wish, as I do, to build a society in which individuals co-operate generously and unselfishly towards a common good, you can expect little help from biological nature. Let us try to teach generosity and altruism, because we are born selfish.' I do not think this is necessarily true. We have played the Prisoner's Dilemma too many times throughout our evolutionary history for selfishness to overcome a longer-term strategy, that of co-operation. Social life, both on the savannah and here in the modern world, is a non-zero sum game. We all benefit if we co-operate, we all lose out if we do not.

Yet we should be aware that co-operation *within* a group is the flip side of aggression *between* groups. The closeness of battle-hardened combat soldiers is a testament to the non-zero-sumness of organized violence. Look after your buddy, because there will come a time when you will need him to look after you. As the stakes get higher, so does the intensity of teamwork and group spirit. The warlike traditions of present-day foraging cultures foster an immense amount of group harmony and selflessness, and no doubt if we could look back half a million years we would see this reflected in the lives of our early human ancestors. War creates unity far more easily than peace could ever do; the legendary Blitz spirit may now be blurred by rose-tinted myth, but there was without doubt a real feeling at the time of national pride and common ambition.

This still seems like a bleak picture of mankind. Co-operation during wartime may have a higher purpose but individual survival is still at the heart of one's motivation. All the words culled from game theory and economics – strategy, co-operation, non-zero sum – paint humans as essentially self-interested. We co-operate so that we can gain more food, sex and status for ourselves, or protect ourselves from predators, environmental hardships or a violent enemy. Are we really so single-minded?

I think not. The discovery in France of that fragile human jawbone, some two hundred thousand years old, shows that early *Homo sapiens* cared for the weak and vulnerable. It was scarred by gum disease, which indicates that the teeth had fallen out, but there was also evidence of bone re-growth in the holes left behind, showing that its owner had survived for at least several months. This individual must have been given either soft bits of food or morsels that had already been chewed by someone else. The owner of the jawbone was unlikely to have been able to reciprocate favours or help with the daily chores. Even if they were somebody's mother or father, the burden of caring for an elderly parent must have been onerous, and kin selection cannot really explain why we care for a parent when they are past reproductive age.

Self-interest may be the defining characteristic of natural selection, but it does not define all of human behaviour; nor, I think, does it give a complete picture of human instinct. In the next chapter I shall try to explore the side of our humanity that eschews survival of the fittest for a kinder, more generous streak. We have concern for others, a sense of fairness; we are affected by feelings of guilt and shame; maybe we even have the capacity for pure altruism. As we shall see, human emotion and value possibly can travel beyond the cold calculus of evolution.

chapter**eight**

Morality and Spirituality – Beyond Instinct?

Abraham Pais was one of the lucky European Jews. He was a brilliant Dutch-Jewish physicist who, after the war, gained a professorship at Princeton and then at Rockefeller University; he died only in 2001. When the Germans marched into Holland in May 1940, the first anti-Jewish measure imposed by the occupying force was a ban on Jews visiting the cinema. At the time of the Nazi invasion, Pais remembered thinking, 'So what? So we can't see movies any more.' Within a year, Dutch Jews were required to have a letter 'J' stamped in their passport, soon after to wear the yellow star, and then a ban was placed on their graduating from a university. By working day and night, Abraham just about completed his Ph.D. in time with a thesis good enough to attract the attention of the great Nils Bohr in Denmark. But Abraham, of course, could not leave Holland voluntarily to visit him. Soon, the Nazi grip tightened and the deportations started, at a trickle initially.

For some time it seemed that high-flying academics such as Pais had a measure of protection, but he distrusted the promises given by the Nazi authorities. Both he and his girlfriend Tineke agreed that he should go into hiding. Tineke was a medical student whom Abraham had met at the beginning of the war while working on his Ph.D. She was not Jewish, a fact that incurred the wrath of

Abraham's orthodox father, but after meeting her he grew to like her, and Abraham's and Tineke's relationship became more serious. She would be instrumental in saving his life.

Tineke persuaded some friends to shelter Abraham in their large house on a canal in central Amsterdam. In the attic they built a hiding place with a false wall. After nine months a Gestapo officer arrived at the house and demanded to search it. Abraham hurriedly climbed into the hiding place and pulled the false wall onto the hole, but it would not fit properly. The German searched the attic, but he did not spot the gap in the wall, or Abraham huddling terrified behind it, watching him. Someone had told the authorities or passed on gossip and betrayed him. Abraham had no choice but to find somewhere else to hide.

Thereafter, Abraham hid with no fewer than nine different families in succession. Tineke collected food coupons for him and borrowed books from public libraries. Abraham rarely ventured outside and spent his time reading, thinking and worrying about his parents. From being a perfectly respectable, law-abiding, fairly well-to-do young man from a good family, he was now a hunted animal with all the instincts to match. Almost the worst part was not knowing how long the occupation would last, or indeed whether it would continue for ever.

In March 1945, after more than two years in hiding, the Gestapo caught Abraham and he was incarcerated – in what state we can only imagine. His initial interrogation lasted thirty-six hours, shows of friendship and empathy alternating with physical threats. He went all that time without food or rest. Of these hours, he wrote, 'Never in my whole life was I more afraid than on the day I went into that prison. The fear was like a physical pain. I couldn't tell you where it hurt, but I remember everything hurting. My body hurt with the pain of fear.' But, almost miraculously, that same week the Allied forces crossed the Rhine and cut off the northern part of Holland. All rail routes to the concentration camps in Germany were cut off and within a month the Germans had retreated. Abraham was released, and days later Holland was liberated by the Canadians.

Abraham's parents also hid in the homes of non-Jews, and they too survived the war, although Abraham's sister and other members of his family were sent to the death camps. By the end of the war four out of every five Dutch Jews were dead. These stories were common across Europe. Of my stepfather's immediate family in the same predicament, thirty-seven members disappeared into the camps. Only one survived Auschwitz, but three survived by hiding with altruistic non-Jews.

None of the many families with whom Abraham stayed were close friends, yet his story is typical of several survivors I have met. Ordinary people put their own lives and the lives of their children in great danger in order to help Jews. They risked betrayal that could lead to their imprisonment or execution. Tineke, too, lived under the constant threat of Gestapo investigations. Those who have read the diary of Anne Frank will be well aware of the huge tensions harbouring Jews caused in these families.

Many of the 'Righteous Gentiles' who helped the Jews in occupied Europe are honoured at the Holocaust Memorial museum, Yad V'Shem, in Jerusalem where more than six thousand trees have been planted in their name. They stand as a testament to the ability of human beings to help others at great cost to themselves, to our capacity for real altruism, indeed sheer heroism. From where did this capacity emerge, and how? It is a fine human quality that appears to be at odds with our selfish genes as well as our self-interested strategies for co-operation and reciprocity.

Appearances, of course, can be deceptive. I find it difficult to believe that the heroism of the Gentile rescuers is merely an illusion. I do not accept the famous cynicism of Beaumarchais when he said that drinking without being thirsty and making love at any time are the only things that distinguish us from other animals. For years, philosophers and pig farmers alike have assumed that humans alone are capable of 'pure' altruism – behaviour that is solely for the benefit of others – and a keenly felt moral sensibility. It is slightly disconcerting, therefore, to find that these 'human' qualities appear to be present in the behaviour of other species.

Altruism in animals

We have already seen how many different animals, even those low down on the evolutionary tree, show some form of altruistic behaviour, sometimes without obvious motives of self-interest. But what of those animals with a brain which is rather more comparable to that of humans and who live in a more advanced society?

Dolphins, for example, are thought to be very intelligent animals, with a large brain relative to their body size. They have a family structure and live in clearly organized groups. They communicate with one another using something akin to language. Dolphins show a number of traits which are reminiscent of human behaviour. For example, the synchronized swimming for which dolphins are so famous is not, as is often supposed, just a beautiful form of travel. When two or more dolphins swim in unison and leap gracefully in the air in identical arcs, these are nearly always new groups of young males, showing off. Very often they are signalling to other dolphins that they are 'the new gang in town' in an aggressive attempt to be a hit with impressionable females. But adult dolphins do clearly care for one another; they will co-operate to ward off a perceived threat rather than just dive for cover. Most impressively, it seems they will gather together to lift a wounded member of the group to the surface of the water so that it can breathe regularly.

Is there evidence of feelings of empathy in some primates? In 1964, Jules Masserman, a famous psychiatrist in Chicago, and his colleagues studied whether rhesus monkeys would forgo food if they knew that by securing the food another monkey would suffer an electric shock. Monkeys were placed in individual cages. Each cage had two levers, one of which when pulled caused a reward of food to be delivered into its cage. However, pulling this lever had an additional effect: an electric shock was delivered to a monkey in the next-door cage. The second lever did not result in any electric shocks being given, but when it was pulled only half the amount of food was delivered. Most monkeys preferred to pull the lever that did not shock their neighbour, even though they were not getting enough to eat. This kind of altruistic behaviour was more likely if the monkey being tested had been cage mates or if they themselves had previously experienced an electric shock. One monkey refrained from eating under such circumstances for twelve days, and generally these rhesus monkeys were almost willing to starve themselves rather than cause another monkey's distress.

It would be unwise to draw too many conclusions from this early experiment. Ideally, it needs to be carefully repeated. But I am confident that now most human experimenters would be very dubious about the ethicality of conducting an experiment of this sort and that more sophisticated, painless research could be carried out in order to evaluate these interesting observations.

Great apes have been seen to show compassion towards members of *another* species. Those who have seen gorillas in captivity will know how frightening they can be, how large and how aggressive to human intrusion into their territory. Binti Jua, a seven-year-old female western lowland gorilla at the Brookfield Zoo in Chicago, who was carrying her own baby at the time, rescued a three-year-old boy who had fallen over the wall into the gorilla enclosure. He dropped eighteen feet onto the concrete floor, hit his head and lay there unconscious. Binti Jua picked up the child, carried him gently and placed him near a door within easy reach of zoo staff. He was then taken to hospital, where he made a speedy recovery.

Bonobos, which are rather rare, are apes closely related to chimpanzees. They are thought to be man's closest living relative. Remarkably, the bonobo was discovered only in 1929; the largest colony live in Zaire. With its long legs and small head atop narrow shoulders, the bonobo has a more delicate build than a chimpanzee. Bonobo lips are reddish in a black face, the ears small and the nostrils almost as wide as a gorilla's. These primates also have a flatter, more open face with a higher forehead than the chimpanzee's, and, to top it all off, long, fine, black hair neatly parted in the middle. They are particularly playful, quite promiscuous, and are more intelligent than either gorillas or chimpanzees.

One bonobo female called Kuni, in Twycross Zoo in Leicestershire, UK, once caught an injured starling. She took the bird and carefully set it on the ground, but it refused to move. Kuni picked it up and gently threw it into the air, but it still would not fly away. Kuni carried the starling to the top of the highest tree in the enclosure. She cautiously extended the bird's wings and then threw it into the air, but still the bird would not fly out of the enclosure. From then on Kuni protected it, especially when a juvenile bonobo started to get curious and would have harmed it.

Frans de Waal and his colleague Jessica Flack are convinced that human morality has deep roots in our primate past. 'While there is no denying that we are creatures of intellect,' they say, 'it is also clear that we are born with powerful inclinations and emotions that bias our thinking and behaviour. It is in this area that many of the continuities with other animals lie.' For de Waal and Flack, the 'proto-moral' behaviour shown by Kuni and other apes forms the building blocks for the evolution of a human moral sensibility. So why are human beings so concerned about other people? Is our much-prized concern for our fellow man just a more sophisticated version of the behaviour observed in Juni and Binti Jua?

Autism and empathy

In order to have a sense of morality we need to understand the state of mind of other people. There is a key moment in the development of children when they begin to grasp that other people have different desires, intentions or beliefs to themselves. A classic experiment involves acting out the following scene in front of a child, usually with puppets. Fred, the first puppet, is inside a room holding a chocolate bar. He hides the chocolate under a cushion, then leaves the room. The second puppet, Annie, comes into the room, takes the chocolate bar out from under the cushion and puts it in her basket. When Fred walks back into the room, the child is asked: where will Fred look for the chocolate? Very young infants think that Fred will know what they know – that the chocolate is in Annie's basket. However, older children, by the age of about four, realize that Fred will look under the cushion. They are beginning to understand that different individuals are capable of making different actions and having different motivations from themselves.

But a few children never manage to make the distinction and will always have great difficulty when it comes to divining the thoughts or feelings of others. They lack what is called a 'theory of mind', an ability to see another person's point of view. Simon Baron-Cohen, clinical psychologist at the University of Cambridge, has called this condition 'mindblindness'. He believes that this is what is missing in the brains of autistic people.

Some autistic people have great insight into their own condition. One young man was reported to say that he was 'always putting his foot in it', that he envied other people's ability to read other people's minds and predict their reactions, thus avoiding upsetting them. Autistic children can understand physical desires like hunger or thirst, and they can make the link from these desires to feelings of sadness, for example, or feelings of pain or discomfort. But these connections are not instinctive, and severely autistic people cannot read the meaning of facial expressions at all. Baron-Cohen cites several examples of high-achieving people who lack this ability; for example, a professor of Mathematics who had won a Field Medal, the mathematics equivalent of a Nobel Prize, could not decode facial expressions from photographs of actors. (Baron-Cohen is developing a system to 'teach' facial expressions to autistic children using a CD-ROM with photographs of actors. They have to be learned, just like the letters of the alphabet or the names of colours.)

If a child points to an object as if to say 'look at that', there is a good chance they are not autistic. By pointing, the child is attempting to divert a person's attention to an object and is recognizing that his or her point of view differs. Gaze direction is an extremely important ability for communicating and understanding others.* From the age of twelve months normally developed babies instinctively follow a person's gaze if it quickly changes direction. Infants will often flick from the adult's eyes to the object the adult is looking at. We have a finely tuned ability to detect whether someone is looking at us, and if not, where in fact they are looking. Remember

* Many years ago, I was having a heavy safe installed by two very burly workmen from Chubb, Ted and Bert. It had to be manhandled two storeys up the fire escape to my office close to the hospital wall, overlooked by the high-security wing of Wormwood Scrubs prison. To make conversation, I said, 'I don't know why I am bothering with a safe really – no-one knows there is any cash here anyway.' Bert just raised an eyebrow, grunted, and cast a glance over his right shoulder.

years ago when your prescient teacher, writing on the blackboard, told you to stop fiddling, that she had 'eyes at the back of her head'?

Autism is a physical syndrome, not a moral failure. Its main component seems to be a lack of empathy. So far, it has been pretty much impervious to any convincing neurological explanation, but a recent discovery may provide a first tantalizing glimpse of the mechanisms behind our capacity for empathy and our 'theory of mind'.

Mirror neurons

Various animal species can learn how to do certain actions by watching other animals first. I joined zoologist Dr Culum Brown of the Department of Animal Behaviour in Cambridge recently to see him experimenting with some very young brown trout. You may not have appreciated that fish have fun watching television; the glass tanks of Brown's baby trout were close to a fifteen-inch television screen on which he was playing a videotape of a slightly older trout eating a red worm. Before seeing this tape, the baby trout had never eaten a red worm; these young trout started to take an interest in the red worms that had been dropping into their tank only after watching this piscine version of the Naked Chef.

Imitation in the primate world is much more sophisticated, as you'd expect. Humans as well as some species of monkey can learn to perform a task – say, hitting a nut with a stone – by watching another person (or ape) perform the same task and copying them. But this process of imitation may give us much more than merely a useful ability to learn how to crack a nut.

Thanks to research in the early 1990s by Giacomo Rizzolatti and his colleagues at the University of Parma, we are beginning to glean insights into some of the underlying foundations of empathy and our theory of mind. Their discovery was the result of a combination of a wonderful moment of serendipity and a keen eye for the significant detail. Initially, Rizzolatti was not looking into the phenomenon of imitation or empathy; instead, he simply wanted to investigate the pattern of brain activity in macaque monkeys when they performed a certain motor task, such as reaching out a hand for food. A neural structure in the frontal lobe called F5, in the pre-motor cortex, is especially active when the monkey is making certain movements such as picking up or biting an object. The researchers wanted to find out if F5 activity changed depending on the size of the object, so they encouraged the monkeys to pick up pieces of apple, raisins, paper clips, and so on, all the while monitoring neural activity in the pre-motor cortex.

In the midst of this experiment the researchers noticed something rather unusual. They realized that F5 was active not only when the monkey was performing the task, but also when the researcher was picking up the object, so as to move it closer to the monkey. Furthermore, the pattern of activity was virtually identical to when the monkey picked up the object for itself. Rizzolatti and his fellow researchers were quick to realize the significance of this activity. After further testing they found that F5 brainwaves were extremely specific to the task in hand: the neurons would fire one way when the researchers were holding pieces of apple in their hand, and another way when they placed them down on a plate. It was as though the monkey was replaying the task in its own mind. Rizzolatti called these structures in the frontal lobe 'mirror neurons'.

Other researchers around the world set out to discover if humans had similar mechanisms in their brains. Firstly, it was determined that when we watch someone gripping an object, for example, muscles in our own hands tense slightly as though we are priming them to perform the same action. Indeed, it is possible to catch oneself twitching, involuntarily, in sympathy with another person's movement. We may make the movement to kick a non-existent ball when watching a player with whom we closely identify attempt to score in a match. When watching a film, have you ever started to duck when a villain throws a punch at Clint Eastwood?

Subsequently, brain imaging studies showed that mirror neurons apparently do exist in the human brain. Vilayanur Ramachandran and colleagues at the University of California at San Diego have conducted an experiment that involves the suppression of certain brainwaves. There are a variety of electrical wave patterns in the brain, of which one type, called mu waves, are found in the motor cortex. These are associated with movement or the intention to move but are blocked when a person moves his or her hand. They discovered that mu waves are also repressed when we watch someone else move his or her hand. The reason for this would appear to be some kind of imitative function. Rizzolatti, too, has conducted experiments that show similar patterns of brain activity between 'doing' and 'watching'.

A relevant part of the brain in humans is Broca's Area, a structure that is used for the production of speech. This discovery was especially intriguing. Rizzolatti proposed that mirror neurons could be the connection between action and communication. Since these structures seem to allow us to interpret and recognize another person's actions, they may be implicated in the development of communication and speech. Perhaps, says Rizzolatti, involuntary twitching in sympathy might have been the first step in terms of gestures and hand movements that finally led to vocalized speech.

Ramachandran is confident that the discovery of mirror neurons in Broca's Area will yield abundant insights into the evolution of the human mind. 'I predict that mirror neurons will do for psychology what DNA did for biology,' he says. 'They will provide a unifying framework and help explain a host of mental abilities that have hitherto remained mysterious and inaccessible to experiments.' He thinks that mirror neurons may be partly responsible for a crucial moment of human evolution called the 'Great Leap Forward'. Somewhere between forty-five and seventy-five thousand years ago there were the beginnings of symbolic art, followed by a significant jump in the complexity of tools and weapons and the invention of rituals such as burying the dead with beads and flowers. The only reason, says Ramachandran, that these cultural advances 'stuck' in the minds and memories of the population was because we had the capability to imitate and understand the actions of others, in the same way that a language can only become lodged in the minds of a population if the bulk of the population has the ability to learn quickly and easily.

But could these mirror neurons have a role beyond that of mentally imitating and then replicating the actions or speech of another person? Could they play a role in a 'deep' form of empathy – putting oneself emotionally in the place of another?

Ramachandran has run the 'mu wave' experiment with autistic children. His preliminary results show that the mu waves are *not* repressed when an autistic child watches someone move a hand. In other words, they may not have the same functioning mirror neurons as the rest of us. This is one enticing explanation of their difficulty in understanding the point of view of others.

If we can replicate the simple movements of another person in our own mind, it follows that we can replicate a sense of pain or pleasure too. Most of us are aware of wincing when we see a person stub his or her toe, or when we watch someone having a tooth pulled out. Some people imagine that doctors get so used to seeing other people in pain that they tend to become less feeling; consequently, it has been suggested that their mirror neurons may become desensitized. There is no evidence for this; in any case, I personally think that doctors do continue to feel, but learn that it's generally not in their patients' interests to show that they feel. But it raises an interesting question about those extraordinary people who are prepared to submit others to torture. It is even more curious to consider that it is frequently said that the most successful torturers are those who pretend to show empathy with their victims, at least for part of the time. What happens to their mirror neurons?

Mirror neurons may well be the key to our ability to understand the emotional state of mind of another person. We have the ability to 'read' other people's minds, we can put ourselves in someone else's place, and to an extent we can understand their experiences, whether pleasurable or painful. It is a talent that starts early in life. After twelve months, babies will start to interpret their environment through the eyes of another; if a parent has a look of fear or disgust when they look at a toy, the child will generally avoid it. Infants as young as three years old will comfort their mothers if they are crying. These actions are the very beginnings of human empathy.

The Ultimatum Game – fairness, guilt, envy

An assumption of game theory, especially as applied to economics, is that rational behaviour is self-interested. Emotions, and particularly emotions such as empathy, are thought to be a hindrance to the real business of looking after number one. In real life, however, rationality does not ignore emotion; the two combine to create complex and fiendishly unpredictable behaviour. This, perhaps, is coming close to the essence of what it means to be human.

The 'Ultimatum Game' is played by two people. One is given some money by the experimenter, say ten pounds. The two players are told that if they agree on how to split the money between them, each will be allowed to keep his or her share. The rules are simple. Player One is told to make a proposal to Player Two on how to split the money. The offered share could be anything between one penny and ten pounds. Player Two can either accept this offer, in which case he or she keeps the cash, or refuse, in which case neither player gets any money. Player Two has no control over the share of the money that is offered and there's no chance to repeat the game.

Rationally, from a purely financial point of view, it makes sense to accept whatever share Player One offers, but in real life, with real people, that is not what happens. If the offer is low, Player Two will often refuse to accept it, thus ruining the chances for both players to keep any money; on average, around 20 per cent of people reject the offer. We can put this down to feelings of spite. Contrary to rational self-interest, they would rather lose the money than be given an inequitable share.

Conversely, self-interest should make Player One choose to offer a low share. He or she must weigh up the chances of Player Two spitefully refusing the offer, but there is no good reason to split the money evenly. Yet in repeated experiments people offer, on average, between 45 and 50 per cent of the money. It is interesting that the average offer, and the likelihood of the offer being refused, does differ from country to country. South Americans, for example, tended to offer less – 35 per cent on average – and were also half as likely to reject the offer.* Even taking the chances of refusal into account, this is far beyond the amount one may think would

* As I write, I sit in a casino hotel in Las Vegas. Here, people seem to be very self-absorbed, less concerned about others than anywhere else on my travels through the USA. I wonder what percentage of the money would be regarded as worth it here?

guarantee Player Two's acceptance of the offer. Players are motivated by more than the desire to maximize their winnings.

This experiment seems to show that we have a sense of fairness predisposing us to share a reward even with a complete stranger. The other side of the coin to a sense of fairness is a feeling of guilt. If we walk away with ninety-five out of a hundred pounds we may feel guilty at exploiting our 'opponent'.

Our sense of fairness is underlined by a stripped-down variant of the Ultimatum Game called the Dictator Game. In this game Player Two has to accept whatever share he or she is given, and Player One, therefore, can divide the money however he or she likes without worrying about losing his or her share. But even then, approximately one sixth of the players divided the money evenly, despite their complete power to 'dictate' the outcome. Our instinct for co-operation brings with it an inclination towards fairness, as well as feelings of anger or spite if someone does not display fairness towards us.

These money games are played out, of course, as if we are rational agents in a laboratory setting. And, of course, the subjects playing these games know they are being watched and their actions recorded. Even under these circumstances, human beings may seem to be morally wanting, but often seem to be fair. We are not automata who maximize our own gain no matter what the situation. And in everyday life all kinds of 'fuzzy' variables come into play. These feelings of fairness and guilt act as extremely useful regulators of our capacity to reciprocate and forge alliances.

You may have all kinds of informal, even unstated, systems of favours with your neighbour. You may, for example, take his children to school every morning, because it is on your way to work. He, on the other hand, is a keen gardener and is happy to weed and water your garden when you are away. No contract has been signed, and there is no hard and fast set of rules surrounding the arrangement; the system will depend on your instinctive feelings of fairness and guilt. If your neighbour is not pulling his weight, ultimately you will discontinue the agreement, even though the cost of giving the children a lift is minimal. Similarly, if you are unable to carry out your side of the bargain you may feel guilt and try to make up for it.

But this is more or less tit for tat. What about altruistic behaviour which does not expect anything in return? Does what I shall call 'pure' altruism really exist? If so, what lies behind it?

Samuel Oliner is the founder of the Altruistic Personality and Prosocial Behaviour Institute. Over many years, his work has investigated altruistic behaviour in, for example, people who rescued Jews in the Second World War, Carnegie Medal awardees and philanthropists. He lists the qualities he believes may be involved:

> Parental role modelling, courage, empathy, learning caring norms, a prevailing moral code that one does not stand by and see another human being perish, self-esteem, social responsibility, self-efficacy, a sense of justice, a feeling that one can make a difference, intrinsic religious factors, agape [fellowship or brotherhood], inclusiveness of others in the sphere of the rescuer's/helper's responsibility, need to help the community, the need for affiliation, self-enhancement, and reduction of guilt.

These, surely, are attributes largely associated with upbringing, i.e. nurture. Do we not return once more to the unresolved arguments between nature and nurture? It is not an easy task to untangle the altruistic impulse, especially when we are dealing with rare and extraordinary acts of kindness or heroism.

Carnegie Medal – showing off?

The Carnegie Medal is awarded to civilians in the United States and Canada who have risked their lives for strangers. Andrew Carnegie established the fund in 1904 after an accident in one of his mines killed 186 workers. He was impressed by the heroic attempts to rescue those trapped in the mine and set up a fund to reward similar acts of bravery. The medal has been given to over eight and a half thousand people and the latest list of recipients is representative. There is a Colorado postman who saw a runaway car containing a seven-year-old boy approaching certain collision at a busy junction. He ran after the car and dived through the open window to pull on the brake. There are four men who almost drowned in their efforts to save a boy from being swept away by strong currents on Lake

Michigan. There is the father and son who ran into a burning house three times to try to save children trapped on the upper floors.

Most of these people had little time to consider the consequences of their actions. They simply jumped into turbulent water or ran into the burning house. To wait and think would have meant serious injury or death for the people in trouble. We would say they acted on instinct, but is it an instinct in the sense that we have used the word throughout this book? Does it stem from a genetic impulse that can be traced back to our evolutionary roots?

An element of heroic behaviour, particularly male heroics, may come down to sex. Women, as we have discussed, may prefer the fireman to the supermarket manager; most of all (it is believed) they just prefer George Clooney in ER. The biology of sexual selection influences all kinds of risky behaviour, and heroic behaviour is no exception. Clearly, the Carnegie Medal winners were not thinking about impressing women, but it would be wrong to ignore the impact of sexual selection and competition; nor should we forget Zahavi's Handicap Principle. Zahavi would say that risky behaviour, especially for the purpose of helping another person, is a perfect method of advertising one's high-quality genes. It is better even than the peacock's tail, because it plays on other people's instincts for empathy. We are attuned to other people's pain and admire those who can put a stop to it. But Zahavi's theory still holds altruism to be essentially self-interested, because it is derived from sexual selection. People get excited by the *sexiness* of heroic behaviour. Hollywood executives, it is said, have a rule about action heroes such as James Bond or Indiana Jones: they dictate that women should want to sleep with him and men should want to be him.

Suddenly, it all sounds rather cynical. Great acts of human kindness have been reduced to an exploitative tactic for enhancing one's attractiveness to the opposite sex. Frans de Waal calls it a profound paradox that 'genetic self-advancement at the expense of others – which is the basic thrust of evolution – has given rise to remarkable capacities for caring and sympathy'. I am not sure it is a paradox, but it is certainly ironic.

The most self-interested drive in human psychology, sex, may be heavily involved in the most selfless acts of heroism. The key word here is 'involved'. Sexual selection does not run the show. In the rich and unpredictable world of human behaviour, no-one is a prisoner of his or her genes. But their presence will always be felt.

We're not all altruists

Most of us harbour a hope that pure altruism, unaffected by any selfish impulse, exists in human life. Nevertheless, we may feel that even the most devoted and humble of hospice volunteers, anonymous donors or committed rescue workers perform good deeds partly because of some feelings of pride, guilt or shame, or perhaps they get a risk-taking thrill.* But we should still admire them, because they fight on the front lines of altruism, and most of us do not.

Kristen Renwick Monroe is Professor of Politics and Associate Director of the Program in Political Psychology at the University of California at Irvine. Her academic life is spent trying to understand moral values and the nature of altruism. Her story of one Carnegie Medal winner, Lucille Babcock, is told in her Pulitzer-nominated book *The Heart of Altruism*.

In July 1987, Lucille, sixty-five at the time and living in Little Rock, Arkansas, was in her office editing the poetry column for her local newspaper when she heard screaming outside. She looked through the window and saw a man dragging a young woman by her ponytail across the road. There was no-one else in sight. Lucille had moderate heart failure which left her breathless and incapable of strenuous physical activity; she also had old injuries to her back and legs which necessitated wearing braces, making her even more immobile. Still, she did not hesitate. Grabbing her cane, she hobbled down two flights of stairs. She describes how she thought her heart would burst through her efforts. The man had already torn the clothes off his victim and was choking her by the time Lucille arrived at the scene. She swung her cane, shouting to the young woman to

* Of course we get job satisfaction out of doing work which helps other people, but we get job satisfaction in all sorts of ways which don't exclude purely altruistic motives at least a portion of the time.

escape, but the man would not stop. Lucille remembered that she felt she might die 'because he was so vicious-looking . . . I'm gonna kill him or he will kill me'. The thug punched her repeatedly, but she kept swinging her cane and shouting. Finally, the man tried to get into his car. Lucille, not knowing whether he had a gun, would not let him go. She slammed the car door on his hand, opened it and hit him again, all the while screaming for help. At last two men heard her screams and held the man down until the police arrived. Lucille was bruised and battered, but she had prevented somebody who was a stranger from suffering serious injuries.

One thing in particular stands out. When she was struggling with the man, she remembers vividly that two women drove past in different cars. Both of these drivers slowed down to look, but once they realized what was going on they accelerated and drove away – away from an elderly woman who was calling for help while being punched and kicked by a man who had clearly been sexually assaulting a half-naked, semi-collapsed, twenty-year-old girl.

Monroe has spent years studying all kinds of 'altruists', from impulsive heroes to long-term philanthropists. She detects some common threads in their feelings towards helping others. Firstly, they do not direct their altruism towards any particular group. There is no interest necessarily in protecting the interests of their 'group' – their fellow nationals, or people of their own race, class or religion. Secondly, Lucille, and other people who have performed similar acts of altruism, say that they 'had no choice'. But, of course, they did have a choice and although they did not necessarily have any rationally considered moral argument, their overriding feeling was to help the person in need. I doubt that this is merely due to instinct.

Lucille's story illustrates that even though most of us are capable of empathy and most of us have feelings of fairness and guilt, pure altruism is less common. Even if we have altruistic intentions, we may lack the courage to see them through. So where altruism exists, it should be treasured. It has its instinctual roots in an ability to understand the pain and pleasure experienced by others; it will be nurtured by our upbringing and our moral environment. Ultimately, though, it is our capacity to combine instinct, emotion and reason that gives us the facility to perform these remarkable acts.

Moral conflict

Humans have attempted to create legal systems since the beginning of recorded history. The remarkable discovery of one of the earliest was made in 1901. A French archaeological team lead by Jean-Vincent Scheil was working in Susa, in Iran – the ancient city, incidentally, where Queen Esther petitioned her king, Ahasuerus (see page 136). During their excavations they found a massive black stone, broken in three pieces, with cuneiform inscriptions. It was the Stone of Hammurabi,* from around 1750 BC, the first recorded legal and moral code. Hammurabi was the Mesopotamian king who established the dominance of Babylon, perhaps the first great metropolis. This huge monolith had ended up at Susa because at some time after 1700 BC it was carried off from Babylon after the city was sacked by the Elamites.

History suggests older legal and moral instructions, but this is the earliest written record we have of a ruler proclaiming an entire body of laws, arranged in orderly groups, so that all might read and know what was required of them. This code of 282 separate laws begins and ends with addresses to the gods. A law code was a subject for prayer, though the prayers here are chiefly curses directed at those who might neglect or destroy the law. Much of the code deals with the organization of society. The judge who blunders in a law case is to be barred for ever, and heavily fined. A false witness is subject to death. If a builder constructs a house which falls down, killing its owner, he is to be slain. If the owner's son is killed, the builder's son is slain too. Indeed, all the more serious crimes are punishable by death. None of these dread statutes – whose morality, the product of modern man's adolescence, we would now undoubtedly question – allows excuses or explanations, with one striking exception. An accused person was allowed to cast himself into 'the river' (the Euphrates – could swimming have been unknown in those times?). If the current bore the defendant to the shore alive he was declared innocent; if he drowned he was guilty. So faith in the justice of the

* The Stone can now be viewed in the Louvre, in Paris.

ruling gods was already established in the minds of men.

One important area was the regulation of family life. The Babylonians punished desertion by a husband or wife and legislated for adultery, adoption of children, inheritance of property and other issues affecting the family. There are many equivalents in virtually all human societies which reflect some of our broad conclusions about how early humans might have behaved. Protecting the integrity of the family unit was extremely important. It meant children were more likely to live through childhood and it cemented the group and kin network.

Hammurabi's code is an outline of what was regarded then as moral. It stressed co-operation between members of society. Negligence, theft, adultery and jealousy were as damaging for the Babylonians as they were for archaic *Homo sapiens*. Virtually every single system of ethics invented since has tried to find a way to provide a good reason for people not to be selfish.

Paul Ehrlich, the distinguished Stanford University biologist who won the prestigious Crafoord Prize of the Royal Swedish Academy of Science, believes that every ethical system originated in the human mind, a biological entity. He does not think, like many dualist philosophers, that there are moral truths out there, waiting to be discovered, that are distinct and independent of the messy mass of neurons that house the human mind. We are bound to our empirical existence, and our moral sense is therefore grounded firmly in the human world.

Ehrlich concedes that the *capacity* to construct a moral system is a product of evolution. We can imagine the consequences of our actions, think about alternatives, and imagine what others are feeling. All these are valuable qualities, and with free will are preconditions for creating a moral system. But the *content* of that system, he believes, is not dependent on our genes. It is an outcome of human culture, and as such can take many different forms. Western parents, for example, believe that punishment of children who misbehave is perfectly reasonable. Inuit parents, on the other hand, would think it highly unethical. Within cultures, too, there are constant moral arguments over extremely important issues – euthanasia, abortion, animal rights, to name the most contentious.

Why, if we all share a common evolved moral sense, do we have no consensus when it comes to these great ethical questions?

Of course, life in a hunter-gatherer encampment did not give us any experiences to answer these particular questions. Such problems only come with civilization, farming and technology. If you're struggling to find enough to eat, no-one has time to worry about mistreating animals. Moreover, Ehrlich says, even with the benefit of several thousand years or more of ethical debate we do not even have a consensus on what the questions are, never mind the answers. Our moral frames of reference can be completely at odds. For utilitarians, the sum of the consequences is important for any particular action. Others may believe in strict moral absolutes that can never be transgressed, such as the taking of a life. Ehrlich uses the example of the tourists who are stranded in a cave by an ocean shore. The water is rising, and the only way out is through a hole that leads upwards to the clifftop. The first tourist, who is unfortunately vastly overweight, gets stuck near the top of the hole. He is out of danger, but the rest of the group cannot escape. Rescuers arrive, and they are faced with a choice: do they blow up the fat man with dynamite, killing him and saving the lives of the others, or do they use drills to free the fat man, which will save his life but seal the fate of his friends because of the time delay? Is there a right answer to this question? Does it depend on whether or not the fat man pleads with them to save his life or begs to be sacrificed?

Apes or angels?

Disraeli thought that Darwin's theory of evolution threatened to put mankind on the side of the apes rather than on the side of the angels, and this horrified him. Disraeli's fears live on. Creationism is the belief in the literal truth of the Book of Genesis, that the world and everything in it, from earthworms to humans, was created in six days some five thousand years ago. Creationists, whose numbers are astonishingly large in the US but seem proportionately less in the UK, are implacably opposed to the ultimate idea proposed by evolutionary theory: that all of us, humans, apes, worms, share a common ancestor. But there is a mass of extremely convincing evidence that supports all aspects of Darwin's theory of evolution, from carbon-dating to the fossil record, from countless experiments in population genetics to computer simulations of natural selection. All the scientific evidence argues strongly that the first few chapters of Genesis cannot be taken as literal truth.

Just as the Creationists, judging from the surprisingly large postbag I get from them, object vigorously to the views of scientists such as me, so there are too many scientists willing to be outraged by the opinions of the Creationists. Just a few weeks before writing this chapter, I was standing in vast open countryside, miles from modern civilization (as I thought), under the hot African sun of the savannah, holding bits of two hominid skulls – one bit from *A. afarensis,* the other from *Homo erectus*. I was about to record a few words to the BBC camera facing me on the nature of instinct and how we had evolved from the two prototypes which I was clutching. Then I was visited by the surreal. My mobile phone rang. It was Richard Dawkins' secretary in Oxford asking if I would join him in signing a letter to *The Times* with other scientists. They were outraged because a junior school in Newcastle had just announced publicly that it was not going to teach evolution in school, only the story of the Creation. I replied that, while I had the greatest respect for Richard Dawkins, I didn't feel I could add my name to the letter. I tried to rescue my credentials by saying I was 'teaching evolution' at that very moment, but I explained that I really didn't believe it mattered so much what this school taught. I doubted it would damage their pupils if it took them a bit longer to find out about what we believe about evolution, if they came to Darwin a little later. What concerned me more was the risk of turning more people away from rational argument, and hence from science, because of a vehement public argument which could not reflect well on the scientists and would not change the view of Creationists.

And, of course, though they may be loath to admit it, scientists themselves are only believers. One of our greatest physicists, William Thompson, Lord Kelvin, President of the Royal Society, affirmed fewer than ten years before the flight of the Wright brothers, 'I can state flatly that heavier-than-air flying machines are impossible.'

Among other pronouncements, he is also on record as saying, 'Radio has no future,' and 'X-rays will prove to be a hoax.'

The certainty of scientists

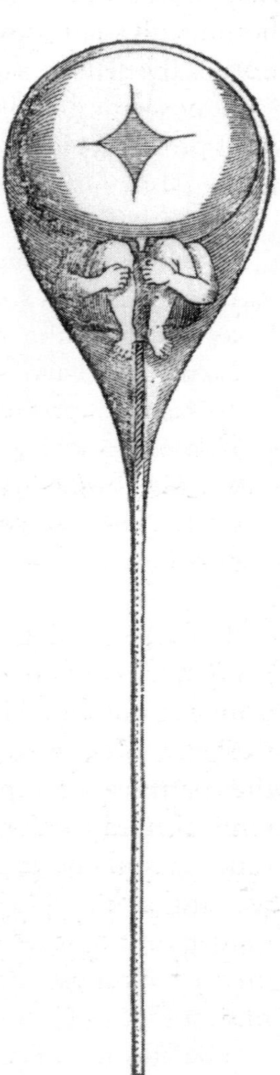

To my mind, our own certainty as scientists has sometimes been a key problem for science in the past, and often still is. We scientists tend to treat our scientific view of nature as a universal truth, and in doing so we often lose what we claim to most pride – our objectivity. When we lose our objectivity our perception of what we think we see is wrongly influenced and can be totally flawed, and the misconceptions can then be passed on to the rest of society.

A good example of what I mean arises from the droplets of seminal fluid examined by Nicolas Hartsoeker, the microscope maker and anatomist. Some time around 1694, he published the famous woodcut in his book *Essay de Dioptrique* which depicts a human sperm. Depicted inside the sperm head is a little homunculus, a tiny man in miniature, in the appropriate fetal position with legs drawn up, and the fontanelle prominently sketched at the front of the skull. The homunculus, a pre-existing human, was envisaged as 'complete' and therefore, of course, had its own gonads and spermatozoa, if a male. These, in turn, contained even more tiny humans, and so on back to the origins of the first human. This notion was enthusiastically endorsed by the famous mathematician Gottfried Leibnitz (1646–1716). He cited Hartsoeker 'and other able men' in support of his own idea that animals, including humans, exist pre-formed and wholly alive 'in miniature in germs before conception'. Essentially, the idea came from an Aristotelian view that a man placed his seed in the woman's womb, where it grew until big enough to be born. So pre-existing ignorance and bias on the part of rather naive observers, combined with the poor optics of the original microscopes, gave rise to what was a totally false observation. This is of great interest because this finding of an homunculus led some ethicists to confirm the ruling that destruction of the semen, say by masturbation, is effectively homicide.

Almost one hundred years later, in 1790, Hartsoeker's observations prompted Rabbi Pinhas Elijah ben Meir to comment as follows in his *Book of the Covenant*, published in Mishnaic Hebrew:

> It has been seen through the viewing instrument, called a microscope, that a drop of a man's sperm, while in its original temperature, contains small creatures in man's image. They live and move within the sperm. Now you can understand how right the Sages were . . . How strange the Talmudic idea that hash-hatat zera [destruction of the seed] is 'like murder' seemed to the 'philosophers' among us before the microscope was invented. They thought that destroying the seed is like destroying wood not yet made into a chair, not knowing that seed is potentially the 'chair' itself, the end product in miniature . . .

The ethics implicit in this statement are impeccable. If the sperm really does contain a little man, with organs complete, his sacrifice is indeed murder. The problem is, of course, that the observation is flawed. Further developments in biology have clearly shown us that the sperm is not a 'person' and therefore wastage or destruction of seminal fluid cannot be seen to be 'like murder'. Our ethical attitude can only be as good as our understanding of the world around us. Just as the Code of Hammurabi now is cruel and morally outdated, it follows that religious or ethical views which are based on a false premise, faulty observation or flawed data are valueless and misleading.

The attitude of scientists who have misplaced certainty is essentially no better than the attitudes of those who profess blind faith in their right to impose their religious views on others, or of those who harass patients attending an abortion clinic in Massachusetts, or of those who used violence to prevent black people marrying with white people because these things were 'against the Bible'. Remember the 'science' of eugenics and its consequences? Scientists who are certain that they are right not only fool themselves, but do greater harm. They are frequently held in respect and awe by many people, and they may poison the wells from which less well-educated people drink.

Among the many appalling threats to our society triggered by the events of 11 September 2001 was one that was almost intangible. It was the violent language which followed the massacre of innocent civilians. I would argue that one of the key causes of 11 September was the violent language of hatred which preceded these murders, language which had slowly dripped corrosively on the minds of the perpetrators and their supporters. Of course, it is difficult even now to understand the enormity of those dreadful deeds or to comprehend precisely what was in the mind of the young men who flew those aircraft into the buildings in New York. To nearly all of us they were murderers, but, paradoxically, to themselves and their followers they were martyrs.

I find it difficult and painful to write what I feel I must now set down. Richard Dawkins is a man whom I greatly admire, whom I think of with fondness and whom I am proud to believe is a friend. He wrote his analysis of the events of 11 September in the *Guardian* under the headline RELIGION'S MISGUIDED MISSILES. He asked where the suicide pilots' motivation and their 'insane courage' came from. He was prompted to ask this question, he says, by his 'deep grief and fierce anger':

> . . . It came from religion. Religion is also, of course, the underlying source of the divisiveness in the Middle East which motivated this deadly weapon in the first place. But that is another story and not my concern here. My concern here is with the weapon itself. To fill the world with religion, or religions of the Abrahamic kind, is like littering the streets with loaded guns. Do not be surprised if they are used.

Was it right to have indulged in this violent language? If only you had not written what seems such a hate-riddled piece. Is this anger truly righteous? Surely, this evil action was no more religious than the action of a deprived Catholic bomber in a café in Londonderry, a madman mowing down seventeen people with a machine gun in a shopping centre in Hungerford, a Hamas bomber killing children at a Tel Aviv bus-stop? Do you not see that 'the religions of an Abrahamic kind' you are always so ready to repeatedly condemn are possibly what formulated for you and the society in which you live the very moral framework you rightly respect so much – to protect life, to uphold justice, to accept human equality, to believe in mercy? Would you not concede that the events of 11 September are more to do with human instinct than any real religious observance, and that possibly what we observed on that day was, more than anything else, the ultimate manifestation of the selfish gene?

Religion

Muir Weissinger, writing in his book *From Calvary to Tokyo*, says, 'Some agnostics and even militant non-believers accept that "faith" in general is an inevitable element in human evolution . . . it is not, it is more like an evolutionary tail that should have "dropped off" many years ago – something that is life-impeding at this stage of human evolution.' To some it will seem sad that such an admirable scientist as Sir Hermann Bondi can endorse this shallow, polemical book with the following words: 'Faith in an absolute unchanging truth is presented [in this book] not only as the arrogant nonsense it is, but also as the great evil and inhumanity . . . a large number of tepid believers who give strength and power to the fanatics and self-serving individuals who lead such a faith system . . .' I speak as one of those 'tepid believers' and perhaps I feel that I am being both patronized and insulted. At least I am reassured by being in pretty good company. It would seem Sir Hermann Bondi's term 'tepid believers' applies to the great majority of people in our sophisticated society. And perhaps we might reflect that the greatest scientists, when speaking away from the topic of their expertise, can make no special claims.

Wade Clark Roof, in his recent book *Spiritual Marketplace: The Baby Boomers and the Remaking of American Religion*, points out that current surveys show that 94 per cent of Americans believe in God, 90 per cent report praying to God regularly and around 90 per cent claim some form of religious affiliation. On the surface, Britain would seem to be substantially different. In 1992, it was estimated that around 14.4 per cent of the adult population of these mostly Christian islands claimed active membership of a Christian Church, though there are considerable variations across the country – notably, Church membership is six times higher in Northern Ireland than in England. But while Church membership is declining, there is a substantial element, as Grace Davie points out in her book *Religion in Britain*, of believing but not belonging. People are now more individual in their religious beliefs and find, perhaps not entirely surprisingly and for reasons well beyond the scope of this book, that belonging to an established Church is no longer entirely relevant to their needs. But religious thinking is still strong and shows little sign of decline. Studies by the European Values Group show that more than 70 per cent of the British and European population believe in God, more than 50 per cent need moments of prayer, and 55 to 60 per cent define themselves as 'a religious person'. Two thirds of Britons believe in the concept of 'sin' and the 'soul', and over 50 per cent in Heaven. At least one third of British and European adults professed to 'often thinking about the meaning of life'. These figures do not show a decline in the last decade or so.

Here is clear evidence that, while religious affiliation may be changing in modern human society, belief in the spiritual continues and shows little decline. Given that we live in a society which is based on rational behaviour, where people are relatively well educated, where there is good communication, where we can see daily evidence of the scientific basis for so much of what we do, it may seem surprising at first that so many people still profess belief in God. Why has such a seemingly irrational belief survived intact?

One possible, indeed likely, solution is that religiosity and hence religion has given *Homo sapiens* an evolutionary advantage. There is also some evidence that religiosity – the ability to feel 'spiritual' – may be inherited. In one study, Dr Tom Bouchard from Minneapolis compared thirty-five sets of identical twins with thirty-seven non-identical twins. Each of the pairs of twins had been brought up from birth by different adopting parents. Identical twins reared apart showed much closer similarity in their religiousness than non-identical twins. If one identical twin was deeply religious, the other was likely to be as well, even if the adopting parents did not have any particular religious tendencies. What was interesting was that a twin brought up in one religion would tend to be as spiritual as its twin brought up in another, even if the parents were agnostic or atheist. So there may well be a genetic tendency towards religiosity, more pronounced in some people than in others, which has long been part of the human condition.

Dr David Sloan Wilson of Binghamton University in New York State is convinced that religion is an adaptation favoured by natural selection. In his recent book *Darwin's Cathedral*, he argues that group selection, so criticized by the neo-Darwinists, was a driving force in promoting religious feelings during evolution. He suggests we should think of society as an organism, the old idea that preceded William Hamilton's revolutionary views. If society is an organism, he asks, can we then think of morality and religion as biologically and culturally evolved adaptations that enable human groups to function as single units rather than as collections of individuals? Dr Wilson brings evidence from groups as diverse as hunter-gatherer societies and urban American congegrations. He suggests that religions have enabled people to achieve by collective action what they never could alone. Ultimately, he is one of those rather uncommon evolutionary biologists who is supportive of religion and religious values. This is a view which may be regarded as scientifically rather unorthodox, but one with which I have a strong sympathy.

The sanctity of human life

One of the stronger arguments that atheists bring against the notion of the existence of God is that no two religions believe in precisely the same values and that each appears to be intolerant of others. If God exists, how can two people believe in opposing views of His or Her manifestation? It is certainly true that each religion has a different moral code, but there are some basic human values, it seems, which are close to being universal. Indeed, Wilson suggests that there is a set of basic human moral values which are seen in widely different societies and which he feels are the products of our evolution. He also argues that 'religion must conform to morality'. Consequently, it seems that in his view morality is not God-given but something that has been arrived at.

A central tenet of the moral code of Western society would seem to be the idea of the sanctity of human life. Of course, the definition of what is life may vary from society to society, from culture to culture, but it is surely likely that there is a selective advantage in such a belief.

As a Jew, I feel that the greatest example of what is close to being a set of universal values is that found in the first five books of the Old Testament. Laid down here are principles which are centred on the sanctity of life. I do not refer only to the Ten Commandments and the divine instruction against murder; the whole structure of Mosaic law, adopted in various guises and in different forms by other Western religions, is centred on this idea. Though it is frequently misrepresented by some of those who would follow the line taken by Richard Dawkins, its ideals have never been transcended. The notion of all men's equality, the clear recognition of the need for justice, the basic idea of individual liberty, the application of a Sabbath – a day of rest which applies to everybody in society, householders, servants, strangers – are essentially about the protection of life and the maintenance of its quality. Respect for life is even shown in this tradition by how domestic animals are afforded protection: rest on the Sabbath, no unnecessary cruelty, care and humanity given to them, even humane methods of slaughter – this is a way of ensuring lack of brutalization and respect for life.

Martyrdom and suicide

The lowest point on Earth is one which supports the least life, and is to humans one of the most inhospitable. The arid beach around the Dead Sea in Israel, with its rocks made of pure potash and salt and its torrid, hot, sulphurous atmosphere, is 1,500 feet below sea level. In those concentrated, undrinkable, oily waters thick with all sorts of minerals, hardly any life is sustained beyond some single-celled green algae by the name of *Dunaliella parva* and primeval bacteria, the red, chlorine-loving Archaea. Whether it was always thus is hard to know, but on this site the Bible tells of the destruction of the cities of Sodom and Gomorrah by fire and brimstone – God's response to the persistent sinning of the people of that valley. Even now, long after that presumably volcanic catastrophe, there is clear evidence that human civilization survived just a few miles away, at least for many centuries. On the west side of the sea's shore, rising well over a thousand feet sheer above this rift valley, are the cliffs of Masada. On this flat plateau there is a natural fortress which was once accessible only by a narrow, steep and dangerous 'snake path'. This archaeological site was where one of King Herod's palaces was built, and it subsequently became the scene of one of the most puzzling and paradoxical events in Roman and Jewish history.

The best account of what apparently happened at Masada is that written by Josephus Flavius in his *The Jewish War*.* Josephus, a Jew from an elite priestly family, was the youthful governor of Galilee, and a rebel fighting against the Roman occupation. Eventually he was captured and he surrendered to Vespasian, who subsequently became emperor; Josephus survived captivity by becoming a Roman citizen, and later a successful historian. Josephus's histories are replete with examples of humans causing their own deaths.

Josephus recounts how, some seventy-five years after Herod's death, a group of Jewish religious zealots overcame the small Roman

* The story of Masada attracted many explorers to the Judaean desert, but it wasn't until 1842 that the site was unequivocally identified. Excavations took place only as recently as 1963 using, among others, many volunteer English students, among whom were a number of my university friends.

garrison of Masada, probably taking it by surprise at night. Until that moment, Masada had always been thought to be totally impregnable and resistant to siege. The narrow path up, the steepness of its cliffs and its many huge water cisterns, buried deep in the rock at the top and providing essential supplies for drinking and agriculture, meant that the inhabitants of Masada could sustain themselves in isolation for long periods. From this vantage point, which today still has a clear view over the whole rift valley and across to the mountains of Jordan, the rebels carried a two-year war to the might of the Roman army, descending from this base at night and harassing their enemy across the surrounding countryside. In the year 73, the Romans finally mustered a force to take effective action. Flavius Silva, commander of the Tenth Legion, surrounded Masada by building eight separate camps and slowly constructed a massive ramp up part of the western walls, possibly using thousands of Jewish slaves. One year later, a battering ram was moved into place, and it became only a matter of time before the defending zealots were defeated.

According to Josephus, some 960 men, women and children were in the fortress. Just two women survived to tell their tale to Josephus; the rest seemed to have decided that, rather than be taken alive, they would cast lots to see who would kill the children and then themselves commit suicide. Josephus records: '[the Romans could do no] other than wonder at the courage of their resolution and at the immovable contempt of death which so great a number of them had shown . . .'

Both Judaism and Christianity, like the main monotheistic religions, regard the sanctity of human life as a central pillar of their moral beliefs. Life is regarded as being given by God. If we are made in the image of God, then it is our duty to protect, maintain and enhance life wherever possible. And consequently Western civilization's morality is based on this one premise. Our moral values stem from our conviction that life is uniquely sacred and the framework of our religious and secular law emanates from that principle. So, given that there is no real evidence that the Romans would have destroyed the entire commune of Masada upon their capture (certainly according to the evidence of the two survivors), we have a paradox. Whether one is a Jewish religious zealot and therefore presumably committed to the sanctity of life, or whether one believes that humans are ultimately most concerned with the genetic imperative of propagation of genes, this action of the besieged on Masada seems extraordinary.

Of course, each religion has its differences. This is always one of the arguments proposed by atheists against the idea of a unifying religious truth, and hence against the idea of there being a God. For Jews, interpretation of the sanctity of life is somewhat different from that in, say, the Christian tradition. According to the Jewish tradition, the actions of the inhabitants of Masada are unusual, because voluntary death was certainly something which was rarely condoned.

In the Old Testament, except perhaps for a vague allusion in the Book of Isaiah and a firmer reference in the Book of Daniel (a strange, apocalyptic book, late in Jewish biblical history),* there is no idea of an after-life. Life emanates from God, who gave it, and death becomes an event of absolute finality. When Job is brought low, from riches to poverty, from health to disease, from having a large family to total bereavement, friendlessness and abject suffering, though he considers death, he never contemplates suicide. In the Jewish tradition, martyrdom starts to be regarded as a 'positive' action only around a hundred to two hundred years before Christ. In Maccabees II, Hanna and her seven sons are subjected to a gruesome death** when they refuse to eat the pork forced on them by the king, Antiochus. All of them profess to a belief in life after death and a conviction that their torturers will suffer future judgement.

From such martyrdom to suicide (under extreme conditions) seems a short step. Rabbi Hanina ben Teradyon, whose martyrdom is remembered every year by Jews celebrating Yom Kippur, the Day of Atonement, was executed at the time of the Emperor Hadrian. In the Talmud, tractate *Avodah Zarah*, it is told how the Romans

* Daniel is possibly the last book to be written in the Hebrew Bible and depicts the sixth century BC captivity of the Jews under Babylon and Persia.

** Scalping, then amputation of the hands and feet, and removal of the tongues, which were fried in a large metal pan.

burned him at the stake with a scroll of the Law wrapped around his body and wool around his heart so that his vital organs would burn slowly and his death would be protracted. As the fire begins to rage, his disciples call to him to 'open your mouth so the fire may enter' to end his suffering. Rabbi Hanina replies, 'Let Him who gave me [life] take it away, for no-one should injure themselves.' The Talmud continues:

> The executioner said to him, 'Rabbi, if I raise the flame and take away the wool from your heart, will you cause me to enter the World to come?' 'Yes,' he replied. 'Then swear to me,' and he swore. Then he [the executioner] raised the flame and took away the tufts of wool from his heart and his soul departed speedily. Then the executioner jumped and threw himself into the fire. And a bat-kol [heavenly voice] exclaimed, 'Rabbi Hanina and the executioner have been assigned a place in the World to come.' When Rabbi [Judah] heard it he wept, saying 'One may attain eternal life in a single hour, another after many years.'

The message here seems to be that Hanina refuses to decide to take his own life, but he does allow someone else, the executioner, to act on his behalf. He brings about his own death only indirectly. Remarkably, the executioner receives approval from a heavenly authority when he achieves eternity by jumping into the flames he himself has created.*

The value of human life is central to Jewish thought. Under Jewish law, there were four categories of execution – stoning, strangulation, decapitation and asphyxiation** – but the respect given to human life was so great that these penalties were hardly ever enacted after the Temple of Solomon had been constructed. The Sanhedrin that enacted a death penalty once in seventy years was called the 'Murderous Sanhedrin'. So many conditions were

* A similar story is told in Talmud tractate *Gittin*, concerning Rabbi Gamaliel's arrest by a Roman soldier, who achieves eternal life by throwing himself off a roof.
** As fully recounted in Genesis Rabbah 65.22, Jakum of Zerototh decides he is so unworthy as to deserve death by all four categories of execution. His ingenuity allowed him to manage this tour de force, and with a fitting end he enters Paradise.

required to be fulfilled by Jewish courts before judicial execution that it wasn't until 1962 that a Jewish court actually recorded condemning a man to death. His name was Adolf Eichmann.

Christianity has an undoubtedly equal recognition of life's sanctity; the interpretation is slightly different, but the basic concept is the same. Jesus himself was a martyr, in a sense a voluntary one. He had the opportunity on more than one occasion to avert his own execution, but ultimately he was crucified and achieved everlasting life. Perhaps partly in consequence, at least until the time of St Augustine, martyrdom continued through Christian history from its earliest days, and martyrs are still revered. A typical account is that of the Christians persecuted by Septimus Severus who in March AD 203 were led into the amphitheatre in Carthage to 'fight' with lions. They chose death rather than compromise their faith. Among them was Vibia Perpetua, just twenty-two years old, with her infant son. Before condemning her, the Roman governor, Hilarianus, implored, 'Have pity on your father's grey head . . . on your infant son.' '*Non facio,*' replied Perpetua. 'Are you a Christian?' said Hilarianus. '*Christiana sum,*' she answered. And it seems that this for Perpetua, as so often before and so often since, was the ultimate religious act. 'It is not with wild animals that I will fight, but with the Devil, but I know I will win victory,' she is reported to have said. Victory for her was life after death.

Nor, in the Christian tradition, is suicide itself necessarily a matter of total condemnation. Matthew recounts his version of the death of Judas Iscariot thus: 'When Judas, his betrayer, saw that he was condemned he repented and brought back the thirty pieces of silver to the chief priests and the elders saying, "I have sinned in betraying innocent blood." They said, "What is that to us? That is your affair." And throwing the pieces of silver down in the Temple he went and hanged himself.' Here, Judas clearly repents, and nowhere does Matthew show disapproval for his action as a suicide. Judas takes his life even though he is not under threat, and this is seen simply as an act of remorse.

It is not until the later tradition, in the writings of St Augustine, that we find criticism of Judas and a strong objection to suicide as an immoral act. Augustine likened martyrs to criminals who had broken the law. He pointed out that death should be avoided, stressing that it was preferable to flee from persecution whenever possible. He also argued that it was wrong for non-Christians to kill themselves. The barbarian invasion of Rome in AD 410 can be cited as an example when non-Christian Roman women killed themselves rather than submit to rape, an act of self-destruction Augustine condemns.

Even in modern times we have condemned suicide. By 1800, many horrific medieval laws had been repealed in England, but attempted suicide was still a crime, its punishment, paradoxically, death. Nicholas Ogarev, writing a letter home to Russia *circa* 1860, described how a man in London who had cut his own throat was revived, tried by the court and then hanged – but his throat wounds broke down and he was able to breathe through the hole in his trachea. After being cut down by the good alderman, they suffocated him. After 1900, attempted suicide in England was punishable by up to two years' imprisonment, and it was only in 1961 that Parliament declared suicide no longer a crime – though, as we have seen recently with cases of voluntary euthanasia, abetting a suicide is still a crime carrying a penalty of up to fourteen years in jail.

Spirituality

One of the most unspeakable accounts of man's absolute depravity, of the most base, perverted and evil instincts, was movingly told by Viktor Frankl, a psychiatrist from Vienna who survived both in body and spirit from the depths of Auschwitz. My very imperfect prose does injustice to the reality he experienced.

He was one of 1,500 'ordinary' people who on one particular train, in the depths of winter, travelled for four days and nights, eighty to each truck, with no room to lie down. All but three hundred of the people on his train were incinerated on the first day of their arrival, arbitrarily chosen by the leisurely movement of an SS officer's finger gently pointing to the left rather than to the right in front of the column of new arrivals. Together with other men, Frankl was stripped naked and all his clothes and possessions removed. He was deloused, his whole body was shaved of all hair, and he was repeatedly beaten.

In the first four days after his arrival, Frankl had just five ounces of bread to eat. He slept in a bunk shared by nine other men, some of them already afflicted with copious diarrhoea, covered by two filthy blankets, his only pillow his soiled and clay-covered shoes. The shirt he was given had to last for six months, he was able to wash only once in many days, and wherever he walked was filth and human ordure. Violent, death-threatening thrashings by the guards for no reason at all were constant, frostbite with loss of toes and swelling of the feet frequent, contact with the corpses of his comrades a regular event, infection and diarrhoea inevitable – but to stop work in the forest where temperatures were often thirty degrees below freezing meant certain execution. Frankl describes how, after the initial reaction of absolute shock, slowly another mood took its place – apathy. After a while, everybody slipped into this blunted and dehumanized state until they became mere 'primitives'. Viktor Frankl describes how one morning 'I heard someone, whom I knew to be brave and dignified, cry like a child because he finally had to go to the snowy marching grounds in his bare feet, as his shoes were too shrunken for him to wear. In those ghastly minutes, I found a bit of comfort: a small piece of bread which I drew out of my pocket and munched with absorbed delight.'

One by one, nearly all Frankl's comrades died of starvation or diseases such as typhus. Their bodies were thrown on carts, and he, among others, was forced to dump the corpses into pits. Others were sent to the chambers. No value was placed on these deaths by any of the guards, and he and his fellow remaining prisoners could no longer feel pity. His own feelings, as he looked about him, he summed up ironically in a quotation from Nietzsche: '*Was mich nicht umbringt, macht mich stärker*' ('That which does not kill me makes me stronger').

What is truly remarkable about Dr Frankl's account – he was one of the very few who survived – is this. He is clear that, after the shock and then the apathy, eventually in spite of the primitive and deprived existence in the camp it was possible for spiritual life to deepen. He says that some were able 'to retreat from their terrible surroundings to a life of inner riches and spiritual freedom'. Only in this way, he says, can one explain the paradox that some prisoners who were not physically robust in themselves survived. He found that, in common with others in the same predicament, he suddenly had a heightened awareness of the beauty of nature. He recounts how one evening when they were lying exhausted on the floor of their hut, the first miserable food of the day in their hands, he and his colleagues saw a fellow prisoner rush in to persuade them to leave the hut to admire an extraordinary sunset. Dr Frankl is convinced that spiritual feelings not only enabled *him* to survive, but all those who walked out of that camp. Hope became a most important emotion for them all.

'They must not lose hope but should keep their courage in the certainty that the hopelessness of our struggle did not detract from its dignity and its meaning. I said [to them] that someone looks down on each of us in difficult hours – a friend, a wife, somebody alive or dead, a God – and he would not expect us to disappoint him.'

Envoi

Perhaps, indeed, as François Voltaire famously wrote in the eighteenth century, 'If God did not exist, it would be necessary to invent Him.' Maybe, with the growth of our huge brain and with the nature of our consciousness, man could not after all stand naked and defenceless on the savannah. Possibly, later in time, with the development of language and the use of symbols, man grew an instinct for spirituality which pushed him to recognize and bury his dead and to believe in a force which shaped his life, and led him to appreciate the preciousness of life in other humans.

Of course, I do not in any way deny evolution, but science does not explain everything, and to pretend that it does seems to me arrogant. Perhaps at some time our beginnings were initiated by a divine force. One thing that is clear to me is that a knowledge of instinct and a view of evolution alone by no means explain our existence, or the way we are. I may well be a poor scientist, but for me, personally, the universe is a most remarkable and beautiful design, one of physical rationality and populated with human creatures possessing insight and a divine intelligence. And for all I know, one of the most remarkable things about our special universe is that it is unique. Of course, it may be that in time the ideas behind quantum cosmology and new universes elsewhere show us that this is not so.

The great Jewish philosopher Moses ben Maimon – Maimonides – who lived from 1135 to 1204, struggled with the notion of creation. The most serious contradiction he faced was that the Bible traditionally teaches that the universe is the result of a pure act of divine creation – creation 'out of nothing'. But Aristotle, who for Maimonides was the most convincing natural scientist, had taught that the universe is eternal, without beginning and without end. Maimonides concluded that had Aristotle clearly proved his theory of the eternity of the universe, one would have to reinterpret the Bible accordingly. Effectively, Maimonides was suggesting that until a better explanation became available the biblical account was the one he intended to follow. Maimonides accepted the authority of biblical tradition in matters of law, but where science was concerned, as a religious rationalist, there were no authorities. What is most rationally convincing here and now, however much it may go against traditionally accepted opinions, is the way we must try to use our God-given intelligence to understand the natural world.

But how can a God exist when there is palpably so much evil in the world? I believe that in spite of our powerful instincts – instincts that, as we have seen throughout this book, influence virtually every aspect of our behaviour – we have, above all, an understanding of good and bad. Central to this belief is the notion of man's free will. Man has the ability and the freedom to choose between what is moral and what is immoral. Yes, I understand that there are rare circumstances – people with a genetic problem such as partial duplication of chromosome 15; people brought up in a grossly deprived environment such as those feral children – where our will cannot be said to be truly free, but for most of us there is basic morality which just possibly is God-given and which is shaped by those aspects of human nature which are divine. How could God exist when there is palpably so much evil in the world? Well, how could it be otherwise? If there truly is free will, the one most powerful explanation of God is that He does not interfere – indeed, cannot interfere. His interference would effectively negate the freedom humans enjoy to do both great good and great harm to one another. Having set our universe in motion, He has to leave it to man to decide how to handle his existence if man is truly free to choose.

So religion does have a purpose. I know that I personally would be far less responsible, far less moral, far less likely to seek the right path were I not to have a set of rules. Often these rules may seem illogical or tedious, but they serve the purpose of discipline, which is necessary for all humans. To be useful and good, religion must conform to morality, a morality I personally consider likely to be divine but which changes as man grows and understands the natural world around him. That morality is critical, because together with religion it gives us a framework to control those emotions that have arisen from the primitive beginnings of life, feelings which are unlearned and inherited – our instincts.

Bibliography

Christopher Badcock, *Evolutionary Psychology: A Critical Introduction*, Polity Press (2000)

Jerome Barkow, Leda Cosmides, John Tooby (eds), *The Adapted Mind: Evolutionary Psychology and the Generation of Culture*, Oxford University Press (1992). Includes Daly's & Wilson's 'The Man Who Mistook His Wife for a Chattel'

Louise Barrett, Robin Dunbar & John Lycett, *Human Evolutionary Psychology*, Palgrave (2002)

Andrew Brown, *The Darwin Wars*, Simon & Schuster (1999)

David M. Buss, *The Evolution of Desire*, Basic Books (1994)

Charles Darwin, *The Origin of Species*, Penguin Books (1982)

Richard Dawkins, *The Selfish Gene*, new edition Oxford University Press (1989), originally 1976

Daniel C. Dennett, *Darwin's Dangerous Idea: Evolution and the Meanings of Life*, Penguin (1995)

Paul Ehrlich, *Human Natures: Genes, Cultures and the Human Prospect*, Shearwater/Island Press (2000)

Dylan Evans and Oscar Zarate, *Introducing Evolutionary Psychology*, Icon Books (1999)

Helen Fisher, *The Anatomy of Love: The Natural History of Monogamy, Adultery, and Divorce*, Norton (1992)

Viktor Frankl, *Man's Search for Meaning*, Washington Square Press (1985)

Clifford Geertz, *The Interpretation of Cultures*, Basic Books (1973)

Susan Greenfield, *The Human Brain*, Weidenfeld & Nicolson (1997)

Richard Leakey, *The Origin of Humankind*, Weidenfeld & Nicolson (1994)

Margaret Mead, *Coming of Age in Samoa*, HarperCollins (2001)

Kristen Renwick Monroe, *The Heart of Altruism*, Princeton University Press (1996)

Steven Pinker, *How the Mind Works*, Allen Lane, The Penguin Press (1998)

Henry Plotkin, *Evolution in Mind*, Allen Lane, The Penguin Press (1997)

Matt Ridley, *The Red Queen: Sex & the Evolution of Human Nature*, Viking (1993)

Hilary and Steven Rose, *Alas, Poor Darwin: Arguments Against Evolutionary Psychology*, Jonathan Cape (2000)

Carl Sagan, *The Dragons of Eden: Speculations on the Evolution of Human Intelligence*, Hodder (1977)

Frans de Waal, *Chimpanzee Politics: Power and Sex Among the Apes*, Johns Hopkins University Press (1989)

David Sloan Wilson, *Darwin's Cathedral*, University of Chicago Press (2002)

Edward O. Wilson, *On Human Nature*, Harvard University Press (1978), reissued Penguin Books (2001)

Robert Wright, *The Moral Animal: Why We Are the Way We Are*, Little, Brown (1995)

Index

NOTE: Page numbers in *italic* refer to diagrams.